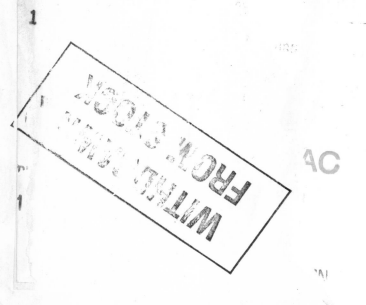

COEFFICIENTS OF NATURAL SELECTION

Biological Sciences

Editor

PROFESSOR A. J. CAIN

M.A., D.PHIL.

*Professor of Zoology
in the University of Liverpool*

COEFFICIENTS OF
NATURAL SELECTION

L. M. Cook

Lecturer in Zoology in
the University of Manchester

HUTCHINSON UNIVERSITY LIBRARY
LONDON

HUTCHINSON & CO (*Publishers*) LTD
178–202 Great Portland Street, London W1

London Melbourne Sydney
Auckland Johannesburg Cape Town
and agencies throughout the world

First published 1971

*This book has been set in Times New Roman and printed in
Great Britain by The Camelot Press Ltd.,
London and Southampton*
ISBN 0 09 104190 2 (cased)
0 09 104191 0 (paper)

288679

To

E. B. FORD

CONTENTS

gene frequency – as a function of gene frequency – examples: inversion frequency in *Drosophila*, frequency-dependent predation, mimicry – reciprocal selection in two species – a general method for finding the stability of an equilibrium

of population size – maximum likelihood method – applied to estimation of
gene frequency – to estimation of selective value

PREFACE

Population genetics is a field in which the theory is much in advance of practical observation and experimentation. The theory was established by three men: R. A. Fisher, J. B. S. Haldane and Sewall Wright. They not only laid the groundwork in the 1920s and early 1930s, but dominated the field for the next twenty or more years so that ideas for which they have not been responsible are only just beginning to make themselves felt. The elementary conclusion from the theoretical system which was developed is that evolutionary change results from change in gene frequency in populations. The ease with which evolution may take place, or the population structure most likely to throw up an evolutionary novelty, depends on the interaction of several systematic and dispersive forces acting on gene frequency. What has been shown in increasingly refined detail is how a particular outcome follows from a particular kind of interaction. But we still know insufficient about the population structure of organisms in nature, and about the size and constancy of selective pressures, to be very clear on the kind of balance that is most general.

The importance of collecting more information from the field is stressed by E. B. Ford in his book *Ecological Genetics*, where he discusses the remarkably great selective forces operating in nature that he has done so much to demonstrate. Writing about 1930, and using the information then available, Haldane stated that the selective pressures acting on the banding types of *Cepaea* differ by amounts of the order of 10^{-5}. Experimental field studies begun by Cain and Sheppard and continued by many others have shown this estimate to be completely wrong: the differences are of the order of several per cent. Earlier, Haldane had estimated the selective advantage of the melanic form of *Biston betularia* in industrial areas to be 50% or more, but at the time few were prepared to believe him. The situation has changed now, but nevertheless a great deal more experimental field work is required. It may be expected to yield new information which is likely to modify the subsequent development of population genetics—the current interest in frequency-dependent selection, for example, stems less from theoretical argument than from direct observation of the behaviour of animals under natural conditions. From the point of view of theory, a general aim should be to get a good idea of the shape of the distribution curve of selective values. We now know that large fitness differences frequently occur, but for statistical reasons small ones are difficult to detect. If we could examine all the loci in a population, would the distribution of selective values be normal, with a mode at selective neutrality, or platycurtic, or bimodal with very few alleles near the neutral point? Assumptions made about the distribution curve have a profound effect on the conclusions to be drawn on such questions as the modification of dominance, the extent of coadaptation and the existence of genetic load, and these in turn affect what can be concluded about such topics as evolutionary strategy. It is therefore essential to learn more about selection in the wild.

It is the aim of this book to give an account for the non-mathematical biologist of the simple theory of selection, and to explain how selective pressures may be measured. Both subjects receive little attention compared to other aspects in texts dealing with the elementary algebra of population genetics. I have tried to relate the theory to well-known examples of selection and natural polymorphism in order to make its significance immediately apparent. Despite the appearance of a few pages, no difficult mathematics is involved, and the mathematical methods are explained as they are introduced. I hope the result allows more people to gain insight from the theoretical literature, as the effort of writing the book has enabled me to do. Most of the material comes from published sources; but in many cases I have modified the symbols used by the original authors in the interests of a consistent notation throughout the book.

No attempt has been made to present a balanced picture of all the forces acting on gene frequency. The effect of factors such as inbreeding and drift has not been touched—they are well dealt with in the textbooks of population genetics. In addition, I have not covered the behaviour of multiallelic and multilocus systems. To do so would be beyond the scope and intention of this book, and recently the algebra has been treated in the excellent account in *Biometrics* by Li (1967).

The first five chapters deal mostly with selective changes of a type that can be discussed without reference to the absolute numbers in a population. This is to oversimplify the picture: there is possibly some effect of density in most selection, even though the effect is small. Similarly, many selective agents that are discussed as if they are completely frequency-independent in effect may well have a frequency-dependent element in their action. In each case, however, it is reasonable to ignore the complicating features while considering the essentials. The subject matter of Chapters 7 and 8 involves the relation between change in gene frequency under selection and population dynamics. A brief review of the concepts involved in discussion of the control of population size is therefore given in Chapter 6.

I am very grateful to all those who have discussed aspects of the book with me. Thanks are due to Professor P. M. Sheppard, Mr Bryan Manly, Drs E. R. Creed, J. D. Currey, P. O'Donald and J. R. G. Turner, and to my colleagues in Manchester, Mr J. G. Blower, Dr R. R. Askew and Dr R. J. Wood, who have read all or part of it. Professor M. Sampford and Mr B. Manly have been kind enough to give valuable help on some of the mathematical aspects. I am particularly indebted to Professor A. J. Cain for his continued interest and valuable advice. Finally, I should like to thank my wife, who was noble enough to take on the typing.

NOTES ON MATHEMATICAL METHODS
AND DEFINITIONS OF TERMS

Morph – a recognizable distinct class of individuals at one stage of development in a population, usually but not necessarily known to be under genetic control.

Gene, locus and *allele* – the terms gene and locus (for genetic locus) have been used interchangeably. In a single diploid individual the locus comprises two alleles (or allelomorphs), which may or may not be identical. Different alleles are different states of the same gene, and are therefore at the same locus. The *gene frequency* in a population is the fraction of all alleles (or twice the number of loci) in a population that are of a certain type.

Selective value – factor describing the difference in success of one kind of individual relative to another as a consequence of selection, usually symbolized by w.

Fitness – used synonymously with selective value.

Selective coefficient – a quantity s such that $(1 - s) = w$.

Net rate of increase – the average number of offspring per individual per generation in a population, here denoted by c.

Intrinsic rate of increase – the natural logarithm of c, always referred to as r.

Stochastic process (*random process*) – one for which only the probability of arriving at a future state can be defined, as distinct from a deterministic one.

Deterministic process – one for which a future state can be defined with certainty. Changes under selection are described here as if they were deterministic. In reality, even in very large populations, the processes are stochastic, and the deterministic equations give the outcome that has the highest probability of occurrence in a large population.

Types of equation – an equation of the form $q_{n+1} = f(q_n)$, where q is, in our case, gene frequency and n is generation, is known as a *recurrence relation*. In words, the frequency in any generation is some function of the frequency in the preceding generation. The equation $\Delta q = q_{n+1} - q_n$ is a *difference equation*. Usually, q_{n+1} is expressed in terms of q_n, so that all gene frequencies refer to the same generation. The equation $dq/dt = f(q)$ is a *differential equation*. In discussing gene frequency the difference equation is usually appropriate when we think of generations as being discrete, and the differential equation when generations are completely overlapping.

Equilibrium – a system is in a state of equilibrium if it remains unchanged once established. The equilibrium gene frequency is referred to as $\hat{q}$. If the system remains at a new point after being disturbed from equilibrium, then the equilibrium is *neutral*; if it returns to the original point it is *stable*, and if it diverges progressively after disturbance it is *unstable*. The Δq equation (see above) is at an equilibrium value of q if $\Delta q = 0$. The equilibrium is *trivial* if $\hat{q} = 0$ or $\hat{q} = 1$, *non-trivial* if $\hat{q}$ lies between 0 and 1, and for our purposes, impossible if it lies outside these limits. It is stable if $d\Delta q/dq$ at $\hat{q}$ is less than zero and greater than -2, unstable if $d\Delta q/dq$ is positive or -2 or less, and neutral when $d\Delta q/dq = 0$.

Sums and products – $\sum\limits_{i=0}^{k} x_i$ means the sum of all values of x from 0 to k.

$\prod\limits_{i=0}^{k} x_i$ means the product of all x's from 0 to k.

Modulus – absolute value of a number irrespective of sign. The modulus of y is written $|y|$.

Differentials and integrals used in the text:

$$\frac{d}{dx}\, uv = u\frac{dv}{dx} + v\frac{du}{dx}$$

$$\frac{d}{dx}\frac{u}{v} = \frac{1}{v^2}\left(v\frac{du}{dx} - u\frac{dv}{dx}\right) \text{ (see following note)}$$

$$\frac{d}{dx}\, e^{ax} = ae^{ax}$$

$$\frac{d}{dx}\log_e u = \frac{1}{u}\left(\frac{du}{dx}\right)$$

$$\int \frac{1}{x}\, dx = \log_e \pm x$$

$$\int \frac{b}{(ax+c)}\, dx = ab\log_e(ax+c)$$

$$\int ba^x\, dx = \frac{ba^x}{\log_e a}$$

Derivative of Δq – If $\Delta q = \frac{u}{w}$ where u and $\bar{w}$ are both functions of q, then $d\Delta q/dq$ may be found as above. When we differentiate Δq at the equilibrium value of q, $u = 0$, so that $\dfrac{d\Delta q}{dq} = \dfrac{1}{\bar{w}}\left(\dfrac{du}{dq}\right)$.

Solution of a second order equation – the two roots of the equation $y = ax^2 + bx + c$ are given by $x = \dfrac{-b \pm \sqrt{(b^2 - 4ac)}}{2a}$. Compare, for example, equation (4) of Chapter 4.

Solution of equations of higher order – the roots are most easily found by graphing to locate the points where the curve cuts the x-axis at $y = 0$. Alternatively, a method of successive approximation may be used (the Newton–Raphson method). If $y = f(x)$, choose a value of x near the required root. Call this x_1. Then $x_1 = x_0 - y\left(\dfrac{dy}{dx}\right)^{-1}$, and x_1 is a better estimate than x_0. Go on until two successive values of x are very close. Equation (7) of Chapter 4 is $6q^3 - 9q^2 + 5q - 1 = 0$. Using this method and starting from $0 \cdot 7$ the subsequent series is $0 \cdot 5787$, $0 \cdot 5096$, $0 \cdot 5000$, $0 \cdot 5000$.

I

SELECTIVE DIFFERENTIALS

Wright (1949) has divided the processes of influencing gene frequency into those which are potentially or actually determinable in magnitude and direction (directional or systematic processes) and those which are potentially determinable in magnitude but not in direction (dispersive processes). He added a third category for any case in which neither the direction nor magnitude can be predicted. Selection may then be defined as 'all systematic modes of change in gene frequency which do not involve physical transformation of the hereditary material (mutation) or introduction from without (immigration)'. An important difference between selection and other systematic forces is that selective pressures tend to zero as gene frequency tends to zero or unity, whereas mutation and immigration rates are not dependent on gene frequency.

When we speak of natural selection we imply that those who survive have a greater fitness than those who do not. In terms of evolution the fittest kind of organism is one which gives rise to progeny thousands or even millions of years hence, and this kind is not necessarily, and perhaps not usually, the one with the greatest relative output over one or two generations. The concept of fitness in evolution has been discussed by several authors (e.g. Thoday, 1953 and elsewhere) and the conflict between long- and short-term requirements is the central consideration in group selection discussed in Chapter 8. Except where specifically qualified we shall define natural selection in a more limited manner throughout this book, as the action of any agency which causes a relative change in the number of progeny from two kinds of organism which survive to reproductive age.

This definition has two limitations. It is restricted to changes taking place over one generation, so that a large selective advantage does not necessarily imply high long-term fitness. Quite apart from the question of relative survival of the two kinds over thousands of years, it does not conveniently include effects extending over two generations, such as the condition of sinistrality in pond snails (Boycott et al., 1931) or the gene *grandchildless* in *Drosophila subobscura* (Suley, 1953), both of which could nevertheless be highly deleterious when they are expressed.

It is also an *a posteriori* definition. We recognize the existence of a selective force only when a change takes place, but a selective agent is one which in future may be expected to produce a change. The position is similar to that encountered when discussing probability. We speak of the probability of an event A, given the condition B. The chance of getting seven heads in a row when tossing a coin, given that the coin is unbiased, is 2^{-7}. The only way to decide whether the coin is in fact unbiased is to determine by practice how often heads turn up, so that the decision is always to some extent subjective. The selective pressure inferred from a change in frequency has similar defects, and selection measured over one generation is not necessarily the most relevant to long-term evolution, but for most purposes the limitations are sensible and practical.

Meaning of selective value

Selective differentials are usually expressed in terms of factors (coefficients or values) which define the relative change in frequency of two forms over a given time. Thus, if a hundred individuals of one type produce 110 progeny while a hundred of another produce only 100 the differential may be expressed as $110/100 = 1\cdot10$ per generation, or as a 10% advantage to the first type. Expressing this in symbols we can call the numbers, or frequencies, of the two types at time 0 A_0 and B_0, and their numbers or frequencies at time 1, A_1 and B_1. Then

$$A_0 : wB_0 = A_1 : B_1 \tag{1}$$

so that

$$w = \frac{A_0 B_1}{A_1 B_0} = 1\cdot1 \tag{2}$$

The cross-product ratio, w, will be known as the selective value or fitness of B, the selective value of A in this case being unity. It does not matter whether equation (1) is in terms entirely of numbers or entirely of frequencies, or if the two sides are expressed in different ways. An alternative procedure is to substitute $(1 - s)$ for w, when s is known as the selective coefficient. In the example, $s = -0\cdot1$ and the selective coefficient of A is zero, so that s is a direct estimate of the relative advantage. The time interval elapsed is not necessarily specified; in the example used it is one generation.

Referring to the definition it will be seen that this treatment is more general than that proposed by Wright: a differential can be expressed in the above manner whether or not the difference between the forms is genetically controlled. As a rule, however, the ultimate aim is to investigate selective differentials acting upon alleles at genetic loci.

When the selective differential is of the order of a few per cent it makes little difference whether it is expressed as the advantage of type *A* over type *B* or the disadvantage of *B* compared to *A*. Indeed the selective value of one type may be expressed as any arbitrary number so long as the other is chosen accordingly to give the correct ratio. Setting one value at unity, however, enables a comparison to be made which is intuitively meaningful. The range of possible values for a disadvantage is from 1, for selective equality, to zero for complete lethality, while the range for an advantage is from 1 to infinity. The selective coefficients therefore vary about zero from 1 to minus infinity. There are two ways of avoiding this asymmetrical distribution. One is to express all selective differences as the disadvantage of the less fit compared with the more fit type. This method has much to recommend it since it provides a value of the kind which is used in ordinary conversation: the per cent disadvantage where 100% indicates lethality. This procedure will be used for the most part, but differentials expressed as an advantage or as the natural logarithm of the selective value, which is symmetrical about zero with limits plus and minus infinity, are frequently to be found in the literature.

The term differential will be used to indicate that there is a difference in the way selection affects two categories of organism. If the difference can be represented in an equation by a factor measuring the relative success of the two categories then this factor is the selective value, or relative fitness, of one category compared to the other. A selective coefficient is a quantity such that (1 − coefficient) = selective value.

The point at which selection acts

In diploid organisms the fusion of haploid gametes initiates each new generation of zygotes, which subsequently grow up to become adults. The cycle of successive reduction division and zygote formation is necessary to maintain the constant chromosome complement from one generation to the next. Although some aquatic animals liberate gametes into the surrounding water, where, in the simplest cases, union is the consequence of chance factors uninfluenced by the parents, the majority of animals have structural mechanisms and behaviour patterns which make gametic union more certain.

Over one generation the gene frequency at a locus may be affected by the relative survival of zygotes bearing different gene combinations, from the time fusion first occurs until the adults have ceased to reproduce, or in special cases, until after the end of reproduction. At reproduction the genotypes may differ in their abilities to produce gametes. For example, an *Aa* individual might generate a greater number of gametes, half of them *a*, than one which was *aa*, because it is more vigorous and robust; or alternatively, one kind of gamete may always arise in excess in the heterozygote, while the homozygotes are equally productive. Having been formed, gametes of different kinds do not necessarily all have the same chance of surviving until fusion can occur. Finally, gene frequency may be altered by mating behaviour which is non-random with respect to the genes or genotypes concerned, or by some mechanism such as differential protection or nurture of zygotes or young by their parents. There is a cycle containing the following transitions:

zygote → maturation to adulthood → gamete formation

→ mating behaviour → new zygote formation

Usually we think of the generation as running from adult to adult —this will be the most frequent convention in the examples which follow because the genes studied are scored in the adult—but more properly the sequence is from one set of newly formed zygotes to the next, when the mating system and the adult stage in one sequence belong to the same generation. The events which take place under random, unselected conditions at mating and new zygote formation are shown in Table 1.2. The different points in the cycle at which selection, or change in gene frequency, can occur are summarized with their causes in Table 1.1. All kinds of selection on autosomal loci can be fitted into this scheme except for selection acting differently on the two sexes, when we have to think of male and female streams from zygote to gamete, coming together at mating (compare Chapter 2).

Equilibrium populations

The most common genetical system to be studied is one consisting of two autosomal alleles, A_1 and A_2 say, at a single locus in a sexually reproducing population. Let the genotype frequencies A_1A_1, A_1A_2 and A_2A_2 be d, $2h$ and r (where $d + 2h + r = 1$), and the gene frequency of A_1 be p, that of A_2 being q (where $p + q = 1$). Now whatever values we give to p and q or to the three genotype frequencies

Table 1.1
Stages in the life cycle at which a change in gene frequency
can occur

stage	category of effect
zygote formation to adult	differential survival
gamete	(a) differential output (b) differential survival
mating	(a) non-random mating behaviour (b) incompatibility between mates other than behavioural (c) incompatibility between progeny and parents

it is always true that $p = d + h$ and $q = h + r$, because the frequency of A_1 consists of all the A_1 homozygotes plus half the heterozygotes expressed as a fraction of the total population, and the frequency of A_2 is the complement of this value. An experiment may be set up in which d and h have any relation to each other which is consistent with the requirement that d, $2h$ and r add up to unity.

Imagine that the experimental population is a very large one in which there are no extraneous factors such as selection or mating preference acting to disturb the random assortment of the pairs of gametes which comprise the zygotes of the next generation. The frequencies of the gametes are $(d + h)$ and $(h + r)$ respectively, so that by the usual rules of compound probability the frequencies of the three genotypes among the progeny are $(d + h)^2$, $2(d + h)(h + r)$ and $(h + r)^2$. But by definition these frequencies are p^2, $2pq$ and q^2. It is easy to see that repeating the process for another generation will not change the relationship of genotype to gene frequency, so that the population is in a state of equilibrium to which it will return if by any means the correspondence between gene and genotype frequency is disturbed.

This is the well-known Hardy–Weinberg equilibrium. Stating the relationship in words we can say that in an infinite population with random mating and no selection, a stable equilibrium state for two alleles will be reached after one generation such that if the gene frequencies are p and q ($p + q = 1$) the genotype frequencies are p^2, $2pq$ and q^2. Proof of the law is given in the books of Li (1955a), Sheppard (1958) and Falconer (1960), and the modest note in which it was stated by G. H. Hardy is to be found reprinted in Peters

Table 1.2
Random mating and the frequencies of two alleles at a locus

d = frequency of $A_1 A_1$ p = frequency of $A_1 = d + h$
$2h$ = frequency of $A_1 A_2$ q = frequency of $A_2 = h + r$
r = frequency of $A_2 A_2$

(a) frequencies of different types of mating resulting from random assortment

females	males		
	$A_1 A_1$	$A_1 A_2$	$A_2 A_2$
$A_1 A_1$	d^2	$2dh$	dr
$A_1 A_2$	$2dh$	$4h^2$	$2hr$
$A_2 A_2$	dr	$2hr$	r^2

(b) frequency and composition of each group of progeny resulting from random mating

type of mating	frequency	type of progeny		
		$A_1 A_1$	$A_1 A_2$	$A_2 A_2$
$A_1 A_1 \times A_1 A_1$	d^2	d^2	—	—
$A_1 A_1 \times A_1 A_2$	$4dh$	$2dh$	$2dh$	—
$A_1 A_1 \times A_2 A_2$	$2dr$	—	$2dr$	—
$A_1 A_2 \times A_1 A_2$	$4h^2$	h^2	$2h^2$	h^2
$A_1 A_2 \times A_2 A_2$	$4hr$	—	$2hr$	$2hr$
$A_2 A_2 \times A_2 A_2$	r^2	—	—	r^2
totals in terms of $d, 2h, r$	1	$(d+h)^2$	$2(d+h)(h+r)$	$(h+r)^2$
in terms of p, q	1	p^2	$2pq$	q^2

(c) random mating, the genotypes of the mates ignored. Frequencies of different kinds of zygotes resulting from random union of gametes

gametes from female	gametes from male	
	A_1	A_2
A_1	p^2	pq
A_2	pq	q^2

(1959). The sequence from random mating of the different classes of adults to the frequencies of progeny which they generate is shown in Table 1.2(a)and(b). Provided mating is at random and selection does not act during the mating stage, the genotype frequencies of progeny may be derived directly from the frequencies of the gametes, as in 1.2(c).

Now the geometric mean of a series of positive numbers cannot be greater than their arithmetic mean. If we take the two terms p^2 and q^2 the geometric mean is pq, while the arithmetic mean is $\frac{1}{2}(p^2 + q^2)$. Consequently, $2pq \leqslant p^2 + q^2$, that is to say, the frequency of heterozygotes is never greater than the total frequency of homozygotes. This inequality may be rearranged so that $2pq \leqslant (p + q)^2 - 2pq$, or $4pq \leqslant (p + q)^2$. But $p + q = 1$, so that $2pq \leqslant \frac{1}{2}$: the maximum frequency of heterozygotes in Hardy–Weinberg equilibrium is one half, a value only reached when $p = q = \frac{1}{2}$. The relation between frequencies of homozygotes and heterozygotes under random mating is shown in Fig. 1.1. Another graphical method of expressing the relation

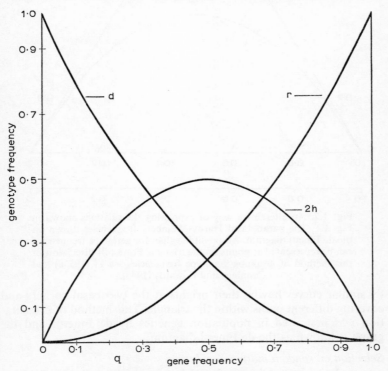

Fig. 1.1. Genotype frequencies for different gene frequencies in a population in Hardy–Weinberg equilibrium.

is to represent the three genotypic frequencies as the perpendicular distances from the sides of an equilateral triangle (Li, 1955a). Equilibrium populations are then described by points lying on a parabola with maximum at $p = q$. Each side of the triangle measures a genotype frequency and perpendiculars from the parabola to the base divide the base in the ratio of gene frequencies (Fig. 1.2). Populations in which the fitnesses are not all equivalent may be represented

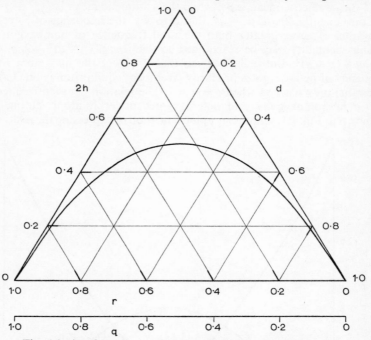

Fig. 1.2. An alternative way of expressing the relations shown in Fig. 1.1. The parabola of Hardy–Weinberg frequencies drawn on the de Finetti diagram. All possible values for genotype frequencies can be represented as points on the triangle. For a full discussion of this method of graphing genotype frequencies see Li (1955a) and Cannings and Edwards (1968).

by similar curves having their origins at the two basal corners and covering different paths within the triangle. This method of presentation was first used in population genetics by de Finetti, and the graph is known as the de Finetti diagram.

Selection on random mating populations

A population is not in Hardy–Weinberg equilibrium if it lies at a point which is off the parabola. An advantage to the heterozygotes compared with the homozygotes will shift the curve of resulting genotype frequencies towards the apex, while a heterozygote disadvantage will depress it towards the base. It should be noted, however, that a position on the equilibrium parabola does not show for certain that the Hardy–Weinberg conditions hold. Indeed, the premise concerning population size cannot be true in practice, and it is possible to imagine selection or non-random mating giving rise to Hardy–Weinberg genotype frequencies. This is so, for example, if there are fitness differences between genotypes, but the selective values of homozygote, heterozygote and second homozygote are in a geometric progression (Li, 1967).

The situations to be discussed are usually ones where combination of gametes takes place at random to form zygotes in frequencies determined by the genotypic frequencies of the parental population; and these frequencies are subsequently changed by selection to form the parental population of the next generation, i.e. there is a zygotic selection. But it is also possible that the gametic frequency differs from the one expected on the assumption of random assortment of parental gametes because there is some kind of gametic selection. The general changes in relative frequency are shown in Table 1.3.

Table 1.3

Changes in frequency of a pair of alleles at an autosomal locus following selection acting on the gametes and the zygotes formed from them

genotype	A_1A_1	A_1A_2	A_2A_2
relative genotype frequencies in parental population	p'^2 :	$2p'q'$:	q'^2
relative frequency of gametes	p'	:	q'
frequencies after selection ($p+q=1$)	$x_1p'(=p)$	:	$x_2q'(=q)$
new parental population after zygotic selection	w_1p^2 :	$2w_2pq$:	w_3q^2
new gametic frequencies	$(w_1p^2 + w_2pq)$	:	$(w_3q^2 + w_2pq)$

Difference equations

It has been seen that a selective differential may be estimated from the change in frequency of one morph relative to another over a period of time. It is often useful to use the selective value to predict what the new frequency will be for a given starting frequency. In the case of selection acting on the gametes (or on two morphs that are not genetically determined) we may represent the selective value of the morph or gamete at frequency p' as x_1, and that of the kind at frequency q' as x_2, as in Table 1.3. Then calling the frequencies before selection p'_0 and q'_0, and the ensuing frequencies after selection p'_1 and q'_1, we may express the new frequency in terms of the old by the equation

$$q'_1 = \frac{x_2 q'_0}{x_1 p'_0 + x_2 q'_0} \tag{3}$$

The changed frequency of q' is expressed as a fraction of the new total. An equation of this form, illustrating one step in a progression, is known as a recurrence relation. Another, general, way of writing the change from one point to the next is

$$q'_1 = q'_0 + \Delta q' \tag{4}$$

so that $\Delta q' = q'_1 - q'_0$ is the change in frequency over one interval of time. This type of equation is called a difference equation. Now substituting the right-hand side of (3) for q'_1 and setting x_1 equal to 1 provides

$$
\begin{aligned}
\Delta q' &= \frac{x_2 q' - q'(p' + x_2 q')}{p' + x_2 q'} \\
&= \frac{-p'q'(1 - x_2)}{p' + x_2 q'}
\end{aligned}
\tag{5}
$$

for any value of q'. Substituting $(1 - s)$ for x_2 we have

$$\Delta q' = \frac{-sp'q'}{1 - sq'} \tag{6}$$

The time interval considered will usually be one generation.

As a rule the system studied is a zygotic one with at least two alleles, when the most convenient value to find is the change in *gene* frequency. If we start with frequencies p^2, $2pq$ and q^2 for the three genotypes A_1A_1, A_1A_2 and A_2A_2 which have fitnesses w_1, w_2 and w_3 the general equation is

$$
\begin{aligned}
\Delta q &= \frac{w_3 q^2 + w_2 pq}{w_1 p^2 + 2w_2 pq + w_3 q^2} - q \\
&= \frac{q}{\bar{w}} \left[\frac{1}{q} (w_3 q^2 + w_2 pq) - \bar{w} \right]
\end{aligned}
\tag{7}
$$

where $\bar{w}$, called the mean fitness of the population by Wright, is equal to $w_1 p^2 + 2w_2 pq + w_3 q^2$. The expression $\dfrac{1}{q}(w_2 q_2 + w_2 pq)$ may be thought of as the average fitness of gene A_2, and denoted by w_j. Then

$$\Delta q = \frac{q}{\bar{w}}(w_j - \bar{w}) \tag{8}$$

that is, the change in gene frequency of gene A_2 is a function of the difference between the mean fitness of the A_2 alleles and the mean fitness for the population as a whole. By writing $w_i = \sum_j w_{ij} Z_{ij}/q_i$ where Z_{ij} is the frequency of ordered genotypes (i.e. genotypes arranged so that the heterozygotes $A_i A_j$ are distinguished for convenience from $A_j A_i$), w_{ij} is the fitness of the ordered genotype and q_i is the relative frequency of the i'th allele, this equation can be shown to hold for any number of alleles (Wright, 1949; Kimura, 1958). It is an important generalization about change in gene frequency under selection, and should be compared with equation (1) of Chapter 8 with which it is algebraically identical. The value of $\bar{w}$ tends to maximize with the result that w_j and $\bar{w}$ tend to converge, thus leading to equilibrium at gene frequencies of zero and one and in some circumstances at a point between as well. The non-trivial equilibrium will be considered in more detail later. For practical purposes it is usually more convenient to find Δq for specific situations, using selective coefficients instead of selective values. The most common examples are given in Table 1.4.

These expressions define one step in the sigmoid progression of change in gene frequency under selection. Each of the curves differs slightly in shape from the others. When using the equations dominance and recessiveness must be taken to refer to the way selection acts on the genotypes in the environment studied, that is, to genotype fitnesses, and not necessarily to their appearances or to the property that is scored. When there is no dominance the change produced by selection on zygotes is half that effected by gametic selection (compare iv and equation 6 in the text). Selection for a recessive character does not produce exactly the same change as selection against a dominant one, as will be seen from equations (i) and (iii), unless the selective coefficients are so small that the denominators tend to unity. Otherwise a coefficient s_1 for a dominant allele is equal to $-s_3/(1-s_3)$ where s_3 is the coefficient of the recessive allele. If a recessive allele is rare in a population it will usually be present in heterozygotes, upon which selection in its favour cannot act. For the same reason selection favouring a dominant will not be so effective when the dominant allele is at a high frequency and the recessive alleles, present in-

Table 1.4

Equations for change in gene frequency under different kinds of selection

condition	initial frequency and fitness			Δq	equilibrium $(\hat{q})$
	A_1A_1 p^2	A_1A_2 $2pq$	A_2A_2 q^2		
(i) complete dominance selection against A_2A_2	1	1	$1-s$	$\dfrac{-sq^2(1-q)}{1-sq^2}$	0 or 1
(ii) complete dominance selection against A_2-	1	$1-s$	$1-s$	$\dfrac{-sq(1-q)^2}{1-2sq+sq^2}$	0 or 1
(iii) complete dominance selection against A_1-	$1-s$	$1-s$	1	$\dfrac{+sq^2(1-q)}{1-s(1-q^2)}$	0 or 1
(iv) heterozygote intermediate	1	$1-\frac{1}{2}s$	$1-s$	$\dfrac{-\frac{1}{2}sq(1-q)}{1-sq}$	0 or 1
(v) heterosis	$1-s_1$	1	$1-s_3$	$\dfrac{+pq(s_1p-s_3q)}{1-s_1p^2-s_3q^2}$	$0,\ \dfrac{s_1}{s_1+s_3}$ or 1
(vi) general	$1-s_1$	$1-h$	$1-s_3$	$\dfrac{+pq[(s_1p-s_3q)-h(1-2q)]}{1-s_1p^2-s_3q^2-2hpq}$	$0,\ \dfrac{h-s_1}{2h-s_1-s_3}$ or 1

creasingly in heterozygotes, cease to be eliminated. The result is that the progress with time of selection for a dominant allele is more rapid than for a recessive at low frequencies but less rapid at high frequencies (Fig. 1.3). At low frequencies there may be a considerable time lag (up to hundreds of generations in the recessive case) before the response becomes noticeable.

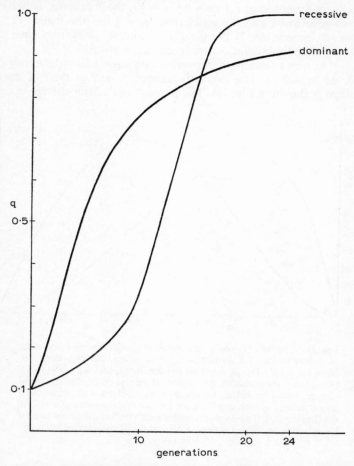

Fig. 1.3. Progress of gene frequency from a starting value of $q = 0.1$ under selection for a dominant and for a recessive character. The fitness values are 1 2 2 for the dominant condition and 1 1 2 for the recessive. The recessive curve requires nearly one hundred generations to move from 0.01 to 0.1, while the dominant one covers the same distance in four generations (compare equations 45 and 46).

32   Coefficients of Natural Selection

The progress of frequencies for a simple situation when one allele is lethal in the homozygote, i.e. $s = 1$, may be considered. If the gene is completely recessive, case (i) in Table 1.4 becomes $\Delta q = -q^2/(1 + q)$ and the frequency in each generation is reduced by this amount to form a harmonic series. Starting from $q_0 = \frac{1}{2}$ the series is $1/2, 1/3, 1/4, 1/5 \cdots 1/(2 + n)$. If the heterozygote is intermediate in fitness (case iv) then $\Delta q = -\frac{1}{2}q$, the frequency is halved in each generation and the series from $q_0 = \frac{1}{2}$ has the value $1/2^{n+1}$ in the nth generation. If the gene is dominant, then from case (ii) $\Delta q = -q$ and elimination occurs in one generation.

Another way to compare the various Δq equations is to graph Δq on q, or to differentiate Δq with respect to q. The type of curve obtained is shown in Fig. 1.4. The point of maximum change in rate

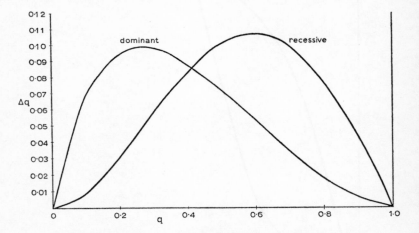

Fig. 1.4. Curves of Δq on q for a dominant and a recessive character as shown in Fig. 1.3. With the fitnesses given, the equations are $\Delta q = q(1 - q)^2/(1 + 2q - q^2)$ for the dominant, and $\Delta q = q^2(1 - q)/(1 + q^2)$ for the recessive. The points of maximum change in rate of change, found by differentiating these expressions with respect to q and equating to zero, are $0 \cdot 24$ for the dominant condition and $0 \cdot 60$ for the recessive. These values compare with $0 \cdot 33$ and $0 \cdot 67$ obtained if the selective coefficient is very small and the denominators are taken to be unity.

of change can be found in each case by equating $d\,\Delta q/dq$ to zero, and if there is balanced polymorphism (expressions v or vi) the type of approach to equilibrium will be found by calculating $d\,\Delta q/dq$ at the equilibrium point (Chapter 4). The point of equilibrium in case (v) occurs where $\Delta q = 0$, from which it follows that $s_1 p - s_3 q = 0$.

Consequently the equilibrium gene frequency

$$\hat{q} = \frac{s_1}{s_1 + s_3} \tag{10}$$

and

$$\frac{\hat{q}}{\hat{p}} = \frac{s_1}{s_3} \tag{11}$$

The expressions for relative gene frequency at equilibrium may therefore be rewritten as follows, substituting s_1 and s_3 for p and q.

Genotype A_1A_1 A_1A_2 A_2A_2

Frequency $s_3^2 +$ $2s_3s_1 +$ $s_1^2 = (s_1 + s_3)^2$

This is sometimes useful when an equation can be simplified if it is written entirely in terms of selective coefficients, and is a convention which has been used particularly by Haldane (1954).

More than one selective agent

It is possible for more than one selective pressure to act on the genes at a locus. If two forces act simultaneously their joint effect may be expressed by summing the Δq equations defining each of them. For example we may consider a situation in the snail *Cepaea nemoralis* where pink is at an advantage over yellow because it is more cryptic and so less subject to predation but at a relative disadvantage because the homozygote is more susceptible to mortality at high temperatures. In colour pink is dominant to yellow, heterozygotes being indistinguishable. The joint effect of both types of selection can be expressed by

$$\Delta q = -\frac{sq(1-q)^2}{1 - 2sq + q^2} - \frac{tq^2(1-q)}{1 - tq^2} \tag{12}$$

where q is the frequency of the pink gene, s is the visual advantage and t the physiological disadvantage. If s and t are small the denominator becomes close to unity and the equation is approximately

$$\Delta q = -sq(1-q)^2 - tq^2(1-q)$$
$$= -pq(sp + tq) \tag{13}$$

Reference to expression (v) shows that if $1 > \hat{q} > 0$, which is only possible if s and t have opposite signs, there is a stable situation arrived at by a balance of selective forces.

It is also possible that the two selective agents act serially rather than simultaneously, when the outcome is different. If we write

$$q_1 = (1 - s)(p_0 q_0 + q_0^2)$$
$$= (1 - s)q_0$$

B

and

$$q_2 = (1 - t)q_1^2$$

assuming small selective coefficients, then

$$\Delta q = (1 - t)[q(1 - s)]^2 - q \tag{14}$$

so that

$$q = 0 \text{ when } (1 - s)(1 - t)^2 = 1/q$$

Other expressions for two-step selective processes may be derived in a similar manner but they quickly become unwieldy. The important point to note is that if several selective forces act on a locus within one generation the outcome will vary depending on whether they act simultaneously or in succession.

An alternative method of expressing the situation described in equation (12) has been used by Cain (Cain and Currey, 1963a) and by Li (1967). Cain and Currey discuss a case in which there is heterosis due to non-visual properties of the genes, and visual selection acting on the recessive homozygote, which for small selective values can be written

$$\Delta q = w_3 cq^2 + pq - q(w_1 p^2 + 2pq + w_3 cq^2)$$

so that the two selective values acting on the homozygote, w_3 and c, are expressed as a product. Setting $s_1 = 1 - w_1$, $s_3 = 1 - w_3$ and $k = 1 - c$ this expression becomes

$$\Delta q = pq(s_1 - s_1 q - kq - s_3 q + s_3 kq) \tag{15}$$

If the relation were described as in equation (12) we should have instead

$$\Delta q = w_3 q^2 + pq - q(w_1 p^2 + 2pq + w_3 q^2)$$
$$+ c'q^2 + pq - q(1 - q^2 + c'q^2)$$

where the w's are assumed to have the same values as before but c is replaced by c'. On putting $(1 - k')$ for c' this simplifies to

$$\Delta q = pq(s_1 - s_1 q - s_3 q - k'q) \tag{16}$$

Equating (15) and (16) we find

$$k' = kw_3 \tag{17}$$

The two procedures are equally useful when used to demonstrate the relation which must exist between k and w_3 in particular situations, and there are circumstances in which one or other is to be preferred. If the aim is to find a numerical value comparable to those discussed before, it is better to use the method of equation (16).

Since it is not always clear what is being measured, it is worth comparing the two methods by means of an analogy. Suppose that a man has the task of assessing the losses suffered by an army fighting under unfavourable conditions. He may know that 15% of all losses are due to enemy action and 20% to disease. In all, 35% of the personnel die, giving an overall chance of surviving of 65%. This is the essence of the argument of equation (16). He also wants to assess the chances of surviving each danger in the absence of the other—probabilities of the kind used in equation (15). First, he imagines all the disease to have struck before the fighting started. The probability of survival is 80%. Fifteen per cent of this fraction, or 18·75%, are then killed in action, so that the survival is the product of these two probabilities, i.e. 65%. He next supposes disease to have struck after the cessation of hostilities. The survival rates are 85% from fighting and 76·5% from disease. The overall survival is 65%. A mean value is obtainable for each probability by saying that the product of the two survival rates E_1 and E_2 is 0·65 and the mortalities $(1 - E_1)$ and $(1 - E_2)$ are in the ratio 15 to 20. The survival rates then come out to be 83·4% for enemy action and 77·9% for disease.

Partitioning fitnesses

There are other circumstances in which it is useful to break up a selective value into components. This is true of the well-known sickle-cell condition in man (Allison, 1955). Certain populations in Africa are polymorphic for an abnormal haemoglobin type—haemoglobin *S*—which differs from the normal molecule in its ability to retain oxygen. Homozygotes for the *S* gene suffer from severe anaemia due to the inefficiency of the erythrocytes as transporters of oxygen. They are also known to be more susceptible than other genotypes to various bacterial diseases (Eeckels *et al.*, 1967). The heterozygotes are little affected, but the erythrocytes, or some other factors associated with the blood stream, provide a worse environment for one of the endemic malarial pathogens than the blood of normal homozygotes.

Where the malaria is endemic both homozygotes have lower fitness than the heterozygote. The difference in fitness between the heterozygote and the sickle cell homozygote is very little influenced by the environment, being due to an 'innate' debilitating condition, although it might well change if people could breathe almost pure oxygen instead of air. The reduced fitness of the typical homozygote, on the other hand, is a function both of the likelihood of surviving malaria and the likelihood of contracting the disease in the first place. A consequence is that the frequency of the gene is much lower in the descendents of West Africans in America than among the contemporary West Africans themselves, because of the absence of the malarial parasite in the New World (although hybridization with American whites must also play a part). The situation would be expressed by writing $(1 - s . v)$ for the fitness of the typical homozygote and $(1 - t)$ for the heterozygote, where s is the coefficient measuring differential survival of the host when infected by the pathogen and v is the likelihood of coming into contact with it. The selective coefficient of the heterozygote t is probably positive, so that when v is zero the heterozygote is at a slight disadvantage owing to the lowered oxygen-carrying capacity of the blood.

In a study of the mutant gene *ebony* in *Drosophila melanogaster* Moree and King (1961) showed the extent to which fitness may be dependent on density. The fitness of the homozygote e/e was 0·915, compared with 1 for the heterozygote, when the density was at its lowest: one fertilized adult female per culture. It fell as the density of adults was increased, whereas the relative fitness of the $+/+$ homozygotes remained almost constant with density. General terms for the relative fitnesses are:

genotype $+/+$ $+/e$ e/e

fitness $(1 - s_1)$ 1 $(1 - s_2 f(N))$

In the terms of Moree and King $s_2 = 1 - V_1$, where V_1 is called the genetic component of viability and taken to be 0·915. The overall viability $V = V_1 \hat{V}$ where $\hat{V}$ is the environmental component of V. Therefore

$$f(N) = (1 - V)/(1 - V_1) \tag{18}$$

At the lowest density $f(N)$ has a value of 1, and the subsequent relation observed between N and $f(N)$ is shown in Fig. 1.5, where it appears that a relatively small increase in the adult density produces a large reduction in fitness, after which further increase has little effect.

Situations where there is selective predation are of general importance. Here the difference in fitness between two types is some sort of compound of the differential due to discriminatory behaviour of the predator and the incidence of predation. We cannot use a simple model, such as for the sickle cell polymorphism, because when the incidence of predation is 1, no differential can occur. The following treatment is adaptable to a number of situations involving predation.

In an experiment designed to test the selective value of mimicry an artificial mimic of a naturally occurring model was created by

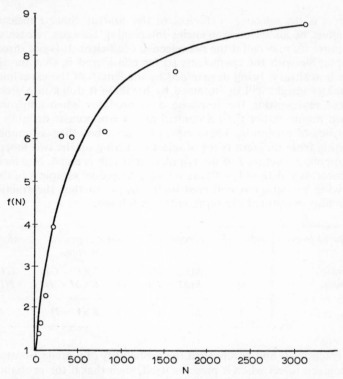

Fig. 1.5. Relation between the environmental component of fitness
$f(N)$ and density for the ebony homozygote of *D. melanogaster*.
Data from Moree and King, 1961.

painting the wings of a large diurnal moth to resemble the pattern
of a distasteful butterfly (Cook, Brower and Alcock, 1969, and
earlier). This is relatively easy to do because the model has a simple
pattern of red and grey spots on a black background, while the moth
is a uniform dark colour over most of the wing area. The artificial
mimics were then released, together with equal numbers of controls
which had black paint marks instead of red and grey, in localities
frequented by the butterfly model. A sample was then recaptured in
assembling traps containing female moths.

Suppose a frequency M_0 of mimics and L_0 of controls was released
($L + M = 1$) and frequencies M_1 and L_1 were recaptured. Then as in
equation (6)

$$\Delta L = \frac{-tLM}{1 - tL} \qquad (19)$$

where t is the selective coefficient of the control. Since t measures the effect of all selective agencies intervening between release and recapture we may call it the net selective coefficient. It is not directly comparable with the coefficients so far considered in that the time scale is arbitrary, being determined by the length of the experiment. A greater insight will be obtained by breaking it down into a component representing the response of a predator when confronted with a mimic rather than a control and a component defining the incidence of predation. There may, of course, be other components resulting from different types of selection acting on the two morphs.

Suppose a fraction I of the population at risk is eaten, of which G are controls and $H(= 1 - G)$ are mimics. A random sample K of those surviving predation is collected in the traps, so that the fractions eaten and recaptured are represented as follows.

ype of insect	released	recaptured	not recaptured not eaten	eaten
control	L	$K(L - GI)$	$(1 - K)(L - GI)$	GI
mimic	M	$K(M - HI)$	$(1 - K)(M - HI)$	HI
	1	$K(1 - I)$	$(1 - K)(1 - I)$	I

The fractions G and H represent the probability of a predator attacking a given insect when it presents itself, such that if the probability of attacking a mimic is 1 the selective coefficient y, which estimates the disadvantage of the control, is given by

$$y = 1 - (H/G)(L/M)$$
$$= (M - H)/M(1 - H) \qquad (20)$$

This expression is similar to those used by Woolf (1955) and Edwards (1965) to measure differential susceptibility to peptic ulcer in two blood group classes, the ratio given by Edwards being the complement of y. In the prese nt example y may be called the instantaneous selective coefficient.

Now the fraction of controls surviving predation is $u = 1 - GI/L$, and the fraction of mimics surviving is $v = 1 - HI/M$ so that

$$(1 - t)L : M = Lu : Mv \text{ (cf. equation 1)}$$

and

$$t = \frac{LI(M - H)}{(M - HI)} \qquad (21)$$

Consequently, from (19) and (21)

$$\Delta L = -\frac{I(M - H)}{1 - I} \tag{22}$$

By rearranging equation (20) it is seen that

$$H = \frac{M(1 - y)}{(1 - My)}$$

which expression may be substituted in (21) to give

$$t = \frac{Iy}{(1 - y)(1 - I) + Ly} \tag{23}$$

This equation shows the relation between t, the coefficient of the conventional type representing the ratio of survivors, y the new coefficient representing the ratio of individuals taken by predators, and I the rate of predation. It is illustrated in Fig. 1.6. When $I = 0.5 = M$ equation (23) reduces to $t = y$. For higher rates of predation t is greater than y and for lower rates is less than y. The coefficient y measures a behavioural response of predators in a choice of situation, which in certain circumstances may be independent of time, whereas t is a coefficient for a discrete time interval. In a case like that of the mimicry experiment y is the most useful factor to find; but if t can be estimated for a period of one generation it is comparable to coefficients found by other means and in some circumstances it may be useful to derive t, knowing y. It will be noticed that since there is a term in L in the denominator of equation (23) the coefficient t depends not only on I and y but also on the initial frequency; that is to say, t is a frequency-dependent coefficient (compare Figs. 1.6a and b), while y is independent. It would be equally possible to define H and G as summing to I, whereupon y would vary with initial frequency. The present definition is biologically meaningful in a wide variety of situations, however, where frequency does not influence the preference of a predator in any way. For example, melanic genes in moths are maintained at a high frequency by predation in many industrial areas. Moth species such as *Biston betularia* and *Gonodontis bidentata* are present as adults for a few weeks of each year, when they rest during the day on trees and other surfaces. It is possible that the principal selection is imposed by a common bird species such as the house sparrow, each individual of which may attack a moth only once or twice in its lifetime. The selection pressure imposed depends on the mean conspicuousness to the birds of the prey morphs on the available backgrounds. In these conditions it

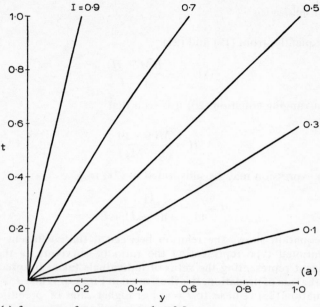

(a) frequency of one prey morph = 0·5

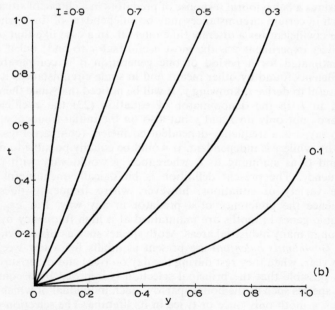

(b) frequency of one prey morph = 0·2

Fig. 1.6. Selective predation acting on a pair of prey morphs. The relation between y, a measure of predator preference, t, the selective coefficient of one prey morph, and I, the intensity of predation.

is quite independent of factors such as relative abundance or frequency, and conspicuousness is most satisfactorily measured by a coefficient of the kind of y. The treatment given here applies equally well to a genetic polymorphism: if there is dominance, q^2 or $(1 - q^2)$ may simply be substituted for M. It must be remembered, however, that if other selective agents intervene along with predation, t is a measure of the joint effect.

Populations with overlapping generations

The foregoing discussion is based on a model where generations do not overlap, so that Δq may be measured between a stage in the life cycle of one generation and the same stage in the life cycle of the next. As a rule it does not violate common sense to assume a different time interval or to estimate the average generation time in a continuous population and write a finite difference equation for the estimated interval, but there are certain differences between the two systems. An account of the populational changes involved in the ensuing sections is given in Chapter 6.

Continuous populations increasing

Suppose first that two types, or species, are increasing in numbers exponentially in the same kind of environment but without competing. For each type we may write

$$\frac{dN_i}{dt} = r_i N_i \quad (i = 1, 2) \tag{24}$$

and

$$N_{it} = N_{io} e^{tr_i}$$

then

$$\frac{N_{it}}{N_{io}} = e^{tr_i} \tag{25}$$

Call this expression c_i when t is the average time for one generation to elapse. Then c_i is the net rate of increase, or output, for the type i. Because fitness under the conditions specified depends on nothing but the rate of increase, we may now find the relative fitness from

$$w = \frac{c_1}{c_2} = e^{r_1 - r_2} \tag{26}$$

so that

$$\log_e w = r_1 - r_2$$

where r_2 is greater than r_1 if the difference is to be expressed as a disadvantage. If r_1 is not equal to r_2 the population will diverge in numbers, so that w will be greater the longer the period of time considered. This treatment is identical with that of equation (2).

A direct transition from the equation in Δq to the differential form has been suggested by Cavalli (1950). He takes the equation

$$\Delta q = \frac{pq(s_1 p - s_2 q)}{1 - s_1 p^2 - s_2 q^2}$$

and substitutes $a_1\, dt$ for s_1 and $a_2\, dt$ for s_2, a_1 and a_2 being the complements of intrinsic rates of increase. We then have

$$dq = \frac{pq(a_1 p - a_2 q)\, dt}{1 - a_1 p^2\, dt - a_2 q^2\, dt}$$

and for small values of a and c

$$\frac{dq}{dt} = pq(a_1 p - a_2 q) \tag{27}$$

Starting from the rates of change in numbers for different genotypes in a population, Kimura (1958) has shown that with the continuous model

$$\frac{dq_i}{dt} = q_i(r_i - \bar{r}) \tag{28}$$

where q_i is the frequency of the i'th allele, r_i is the average instantaneous rate of increase for the genotypes carrying i genes and $\bar{r}$ is the average output of the population. They are calculated from

$$r_i = \sum r_{ij} Z_{ij}/q_i$$

Z being the frequency of the ordered genotypes as before, and

$$\bar{r} = \sum r_{ij} Z_{ij}$$

Equations (27) and (28) may be shown to be identical when $1 - a_i$ is substituted for r_i. The equation for non-overlapping generations comparable to equation (28) is (8) viz:

$$\Delta q_i = \frac{q_i}{\bar{w}}(w_i - \bar{w}) \tag{29}$$

If for w we write $1 + s\,\Delta t$, t denoting intervals of time, equation (29) may be rearranged to become

$$\frac{\Delta q_i}{\Delta t} = \frac{q_i(s - \bar{s})}{1 + \Delta t} \tag{30}$$

As Δt tends to zero, equation (30) tends to identity with (28) but with r_i and $\bar{r}$ replaced by s_i and $\bar{s}$. As Kimura points out this shows that there is a correspondence between s_i and r_i, in particular that factors used in discrete models are best expressed as exponents when infinitesimal time intervals are concerned. The relationship is shown in Fig. 1.7. A point considered in connection with population growth

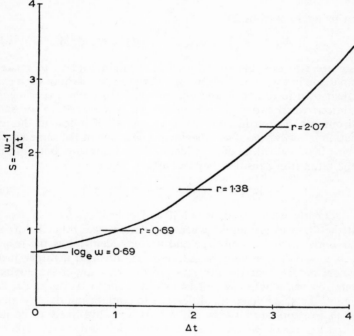

Fig. 1.7. Relation between $s = \dfrac{w-1}{\Delta t}$ and r for two populations having net rates of increase of unity ($r = 0$) and two ($r = \log_e 2$). $w = c_2/c_1 = 2$ when $\Delta t = 1$. Values of t are expressed in intervals of one generation. It is apparent that $s \to r$ as $\Delta t \to 0$.

is illustrated by this graph, namely that a rate measured in terms of r for a given time interval can be converted into the rate for a different unit of time simply by multiplication, whereas one expressed in terms of c cannot. In the example, when $\Delta t = x$, r is equal to $0.69x$ but the new value of s is not so easily found. It will be clear, however, that to produce fitness values concordant with those already used we must work in terms of c.

Populations exhibiting logistic growth

If two populations grow in the logistic manner and occupy different environments or grow together without interaction they may be represented by

$$\frac{dN_i}{dt} = N_i r_i \left(1 - \frac{N_i}{K_i}\right) \quad (i = 1, 2) \tag{31}$$

Then by analogy with (26)

$$\log_e w = r_1(1 - N_1 K_1^{-1}) - r_2(1 - N_2 K_2^{-1}) \tag{32}$$

There are two possible situations which might lead us to make a comparison in this way. They are (a) when the populations each exist on their own in different environments and (b) under experimental conditions when two types are introduced successively to the same environment. The artificiality of this measure of fitness in the first situation is easily seen if we imagine populations of the same species (having identical values of r) in identical environments but differing in the saturation densities they will attain. Then

$$\log_e w = r(N_2 K_2^{-1} - N_1 K_1^{-1}) \tag{33}$$

Consequently when $r \neq 0$, $w = 1$ if and only if $N_2 K_1 = N_1 K_2$, so that the fitnesses are unlike when population sizes differ below the asymptotic levels but converge and become identical as the respective values of K are approached. We see, albeit in a rather metaphorical model, that the fitnesses are different under lax environmental pressures when we might expect them to be the same, and they are the same when the difference in equilibrium density might lead us to suppose that they differed. It is always likely to be misleading to make comparisons between populations in different habitats. This point is discussed by Cain and Sheppard (1956) and Li (1955b). The difficulty with the second situation in which the comparison could be made is the practical one of ensuring that the environment is in fact identical during the two trials.

If the two types are living together in the same environment there will be some sort of competitive interaction between them. This may be expressed in the Gausian manner (Gause, 1934) by the competition factors α and β whereupon

$$\log_e w = r_1 K_1^{-1}(K_1 - N_1 - \alpha N_2) - r_2 K_2^{-1}(K_2 N_2 - \beta N_1) \tag{34}$$

If $r_1 = r_2$ and $N_1 + N_2 = K_1 = K_2$ then

$$\log_e w = rK^{-1}(K - \alpha N_2 - \beta N_1) \tag{35}$$

It should be remembered that α and β are not necessarily constant, but may change with time or density and that their biological meaning is not always clear. In general it would be preferable to find w by measuring numbers in the two types at an interval equivalent to one generation and to use these estimates in equation (2).

Relative fitness from life table data

As a rule populations with overlapping generations, like those with discrete generations, do not grow exponentially nor according to simple logistic rules. They are most usually studied while in some sort of equilibrium with their environment so that they are not increasing in numbers at all. The data from which fitnesses could be calculated take the form of life tables, which provide a picture of the mean survival with age and the age specific birth rate.

The basic calculation is to find w from the ratio of two estimates of c, where

$$c = \frac{1}{l_0} \int_0^\infty l_x m_x \, dx$$

$l_x/l_0 = p_x$ being the probability of an individual surviving to age x, and m_x being the number of offspring produced between x and $x + 1$. The construction of life and fertility tables is described by Leslie and Ransom (1940), Deevey (1947) and Southwood (1966) and the genetical conclusions from them were first discussed by Fisher (1930).

Since the information on output and survival is compiled in the tables for discrete intervals x, the net reproductive value is found by summation rather than integration, i.e.

$$c = \frac{1}{l_0} \sum_0^\infty l_x m_x \tag{36}$$

In the special case where the average output per interval of time is constant and the same for the two types being compared m_x may be ignored so that

$$c = \frac{1}{l_0} \sum_0^\infty l_x \tag{37}$$

It is unlikely that this circumstance often occurs but quite commonly no satisfactory evidence on changes in m_x can be marshalled so that the simplest approximation is to assume it to be constant. But

$$c = \sum_0^\infty p_x = e_0 \tag{38}$$

where e_0 is the mean expectation of life at birth. Thus, under these conditions the relative fitness of two types is the ratio of their expectations of life.

If the mortality d_x per unit time is also constant (but different in the two types) another relation may be seen. The mortality between x and $x + 1$ is the number of individuals which die expressed as a fraction of those alive at x, that is

$$d_x = \frac{1}{l_x}(l_x - l_{x+1}) \tag{39}$$

Substituting in (37) with d constant for all x we have

$$c = \frac{1}{l_0} \sum (l_x - l_{x+1}) \frac{1}{d}$$

$$= \frac{1}{l_0} \cdot \frac{l_0}{d}$$

$$= 1/d \tag{40}$$

Consequently,

$$w = c_1/c_2 = d_2/d_1 \tag{41}$$

When d is large for the interval of time measured (say greater than 0·2) a more accurate estimation of the expectation of life is required. For a constant rate of mortality the equation $l_x = l_0 s^x$ describes the relation between numbers and time, where $s = 1 - d$ is the survival per unit time. The expectation of life is the area of this curve, or $l_0 \int_0^\infty s^x \, dx$. With l_0 taken to be unity this is $s^x/\log_e s + A$. Since $s^x \log_e s$ tends to zero with increase in x and $s^0/\log_e s + A = 0$ it will be seen that

$$A = e_0 = -1/\log_e s$$

The two estimates converge when the mortality in the interval adopted is very small, but with field data it is often impossible to sample frequently enough for this condition to be met.

The method of estimating relative fitness from expectations of life is used by Clarke and Sheppard (1966) to measure the differential predation acting on melanic and typical forms of the peppered moth, *Biston betularia*. By means of a marked release-recapture experiment in an industrially polluted area of Liverpool they estimated the daily survival rate of typicals to be about 20% and that of *carbonaria* individuals to be around 48%. To obtain the selective disadvantage of typicals we therefore have the values $d_1 = 0.8$ and $d_2 = 0.52$, so that $w = d_2/d_1 = 0.65$ and the disadvantage is 35%. The processes of emergence and copulation by newly emerged adults take time, however, so that females do not begin to lay eggs until about one day has elapsed. This day does not directly contribute to the numbers in the next generation: if an animal has a short life span the loss of a day's egg laying is more serious than if the life span is longer. The estimate of w should therefore be adjusted by taking a day from each expectation of life (e_1 and e_2). Now, $w = (e_1 - 1)/(e_2 - 1) = 0.27$ and the selective disadvantage of the typicals is estimated as 73%. Clarke and Sheppard discuss the confidence intervals for this estimate.

We are here dealing with an example in which survival rates are low. For the typicals the expectation of life of 1.25 estimated from $1/d_1$ must be too high if 80% of individuals have died by the end of the first day. The continuous estimation provides $e_1 = -1/\log_e s_1 = 0.62$, and $e_2 = 1.36$. We therefore have $w = 0.46$, suggesting a larger disadvantage of 54% compared with the unadjusted value of 35%. To modify the value for expectation of life, taking account of the absence of eggs on the first day, we have to find the integral from 1 to ∞ of s_i^x, which is $-s_i/\log_e s_i$. This provides $e_1 = 0.124$, $e_2 = 0.654$ and $w = 0.19$. The estimated selective disadvantage of the typicals is 81%.

Clarke and Sheppard emphasize that the selection refers to effects of predation only, and that it depends on the precise conditions of predation existing at the time. These rates of mortality would probably continue to hold throughout adult life, so that the ratio of mortalities is a reasonable estimate of the relative contributions to the next generation. If they changed, however, the calculated values could be very misleading even though the relative visibility of the morphs to predators remained constant. To take an extreme example, suppose that 1000 individuals each of two morphs sustain mortalities of 20 and 30 individuals respectively during a period of study, which is pre-reproductive. Extrapolating from this period of time the expectations of life are 50 and 33 units respectively, and the fitness of the second relative to the first is 33/50, or 0.67. Now suppose that no more individuals die than those observed in the study until all offspring have been produced, and that the output per individuals is the same for each morph. The true relative fitness is the relative frequency of survivors, i.e. 97/98 or 0.99. In fact, the first value is an estimate of $(1 - y)$, and it may be converted into the second by equation (23)

$$t = Iy/[(1 - y)(1 - I) + Ly]$$

where I is the mortality sustained by both morphs, or 50/2000, and

L is equal to 0·5. A discrepancy so extreme between the true value and an estimate of it is not likely to be encountered, but it must be remembered that assessment of selective pressure from a period of predation can only be representative if the predation rate as well as the discriminatory behaviour of the predators are representative. In experiments under semi-artificial conditions the rate may differ quite markedly from that encountered under completely natural circumstances.

Change over several generations

It is sometimes required to obtain from a change in frequency over several generations the mean selective coefficient responsible for the change, or from a knowledge of the selective coefficient the time required for the change to come about. The most direct way to obtain a solution is to substitute successive values in the recurrence equation, or to sum the Δq equation, in order to find the time for a specified change with a given set of fitnesses, and to find fitnesses by trial and error in successive runs. Now that computers are freely available these are simple practical operations, and usually the best course to adopt. An alternative mathematical approach is to integrate the appropriate difference equation and solve for s or n.

The procedure usually recommended is to regard the Δq equation as a differential one, and to integrate after separation of the variables where this is possible. The method may be illustrated with respect to a differential equation using the simple growth equation $dN/dt = rN$ where r is the value over one generation. This may be rewritten

$$\frac{dN}{N} = r\,dt$$

If we wish to find the change in a given time, or the intrinsic rate of increase required to produce a specific change in that time, we have to integrate both sides between the intervals required, thus

$$\int \frac{dN}{N} = \int_0^T r\,dt$$

where T is the time interval. Now the integral of $1/N$ is $\log_e N + K$, so that integrating between 0 and T we have

$$\log_e N_T - \log_e N_0 = rT$$

or

$$rT = \log_e\left(\frac{N_T}{N_0}\right)$$

Since r is defined so that the generation time is unity we can substitute the number of generations n for T, so that

$$rn = \log_e\left(\frac{N_n}{N_0}\right)$$

where n is a number of some discrete units such as generations.

A simple genetic example is the meiotic driving system in the mosquito *Aedes aegypti* (Wood, 1961). In some circumstances males with Y chromosomes carrying the factor D produce spermatozoa which are X and Y in the ratio 1 : 2. The normal allele d gives rise to a 1 : 1 ratio of X and Y spermatozoa. If x is the frequency of female chromosomes produced by D males, y is the frequency produced by d males and the frequency of D in Y chromosomes is $p = (1 - q)$, then:

$$\Delta p = \frac{-pq(x - y)}{1 - px - qy} \tag{42}$$

which, for the particular numerical values given, reduces to

$$\Delta p = \frac{p(1 - p)}{3 + p} \tag{43}$$

It is required to find how many generations have to elapse before fixation occurs of a new mutant D arising in a d population. If the population were infinite the frequency of d would decrease rapidly at first, slowing in rate of loss until eventually the small residue tends asymptotically to zero. Since numbers are finite we may define the point of fixation as the point at which only one d chromosome is left. Equation (43) may be written approximately as

$$\frac{dp}{dt} = \frac{p(1 - p)}{3 + p}$$

when t is measured in generations. The time required for a given change in frequency may be found by integrating over n generations. Thus,

$$\int_{p_0}^{p_n} \frac{dp}{\left[\dfrac{p(1 - p)}{3 + p}\right]} = \int_{p_0}^{p_n} dt$$

so that

$$\int_{p_0}^{p_n} \frac{3 + p}{p(1 - p)} = n$$

Dividing up the left-hand expression we have

$$\int_{p_0}^{p_n} \frac{3}{p} + \frac{4}{1-p} = n$$

Now $\int \dfrac{b}{(ax+c)}\, dx = ab \log_e(ax+c) + K$,

so that,

$$n = 3 \log_e p_n - 4 \log_e(1-p_n) - 3 \log_e p_0 + 4 \log_e(1-p_0)$$

$$= 3 \log_e(p_n/p_0) + 4 \log_e\left(\frac{1-p_0}{1-p_n}\right)$$

If we wish to find the time from formation of a single mutation until only a single non-mutant is left, $p_n = 1 - p_0$, so that $n = 7 \log_e(p_n/p_0)$ approximately.

A similar procedure is followed for autosomal genes and selective coefficients s_i. Li (1955a) has considered the recessive case (case i) with $sq^2 \to 0$, and shown that for the equation

$$\frac{dq}{dt} = -sq^2(1-q)$$

the integrated equation is

$$sn = \frac{q_0 - q_n}{q_0 q_n} + \log_e\left[\frac{q_0(1-q_n)}{q_n(1-q_0)}\right] \tag{44}$$

If sq^2 is not negligible the denominator $(1 - sq^2\, dt)$ has to be used and the equation becomes (M. G. Bulmer in Clarke and Murray, 962),

$$s\left[n + \log_e \frac{1-q_n}{1-q_0}\right] = \frac{q_0 - q_n}{q_0 q_n} + \log_e\left[\frac{q_0(1-q_n)}{q_n(1-q_0)}\right] \tag{45}$$

The equivalent equation given by Clarke and Murray for a dominant selective coefficient (case iii) is

$$s\left[n + \log_e\left(\frac{q_n}{q_0}\right) - \frac{q_0 - q_n}{q_0 q_n}\right] = \frac{q_n - q_0}{q_0 q_n} + \log_e\left[\frac{q_n(1-q_0)}{q_0(1-q_n)}\right] \tag{46}$$

These authors also provide variances for the estimations of s. It is, of course, possible to find n from these equations for a specified value of s; an example and discussion are given by Li.

2

SELECTION ON SEX-LINKED GENES

A sex-linked locus behaves differently from an autosomal one because one of the chromosomes involved determines the sex, usually but not always through a system where one sex is homogametic XX and the other heterogametic XY. The behaviour of sex-linked alleles in the absence of selection is thoroughly treated by Li (1955a) and Falconer (1960).

If there are two alleles at a locus, A_1 and A_2, then in the homogametic sex, which we may for convenience call female, there are three genotypes, A_1A_1, A_1A_2, and A_2A_2. In the males there are only two possibilities: A_1 and A_2. Thus, it is as if we are dealing with genotype frequencies in the female but gametic or gene frequencies in the male. As for autosomal genes in Hardy–Weinberg equilibrium, we may call the frequency of A_1 in males p, and the frequency of A_2 in males q. In the females the three genotypes are in the ratio $p^2 : 2pq : q^2$. If mating is at random and males and females equally common, the matings and frequencies of offspring will be those in Table 2.1. From it we can see that (a) half the progeny are male and half female, (b) the genotypes of the female and the gametic types of the male progeny are in the same ratio as their parents, so that the system is in equilibrium, and (c) the gene frequency of the females is equal to the gene frequency of the males. The difference between this and the Hardy–Weinberg equilibrium for autosomal genes is that while the autosomal equilibrium is reached after one generation the sex-linked one is approached progressively over many generations.

When there is no selection and gene frequency differs between the two sexes, progress towards equilibrium follows a characteristic crosswise pattern. In each generation the male progeny receive all their sex-linked genes from their mothers, so that $q_{1m} = q_{0f}$. The female progeny receive half from their female and half from their

Table 2.1
Types of mating and frequencies of offspring for a sex-linked
gene at equilibrium with no selection

p = frequency of A_1 in gametes of both sexes
q = frequency of A_2 in gametes of both sexes

	random mating	
females	*males*	
	A_1Y	A_2Y
A_1A_1	p^3	p^2q
A_1A_2	$2p^2q$	$2pq^2$
A_2A_2	pq^2	q^3

type of mating	*frequency*	*type of progeny*				
		A_1A_1	A_1A_2	A_2A_2	A_1Y	A_2Y
$A_1A_1 \times A_1Y$	p^3	$\frac{1}{2}p^3$	—	—	$\frac{1}{2}p^3$	—
$A_1A_2 \times A_1Y$	$2p^2q$	$\frac{1}{2}p^2q$	$\frac{1}{2}p^2q$	—	$\frac{1}{2}p^2q$	$\frac{1}{2}p^2q$
$A_2A_2 \times A_1Y$	pq^2	—	$\frac{1}{2}pq^2$	—	—	$\frac{1}{2}pq^2$
$A_1A_1 \times A_2Y$	p^2q	—	$\frac{1}{2}p^2q$	—	$\frac{1}{2}p^2q$	—
$A_1A_2 \times A_2Y$	$2pq^2$	—	$\frac{1}{2}pq^2$	$\frac{1}{2}pq^2$	$\frac{1}{2}pq^2$	$\frac{1}{2}pq^2$
$A_2A_2 \times A_2Y$	q^3	—	—	$\frac{1}{2}q^3$	—	$\frac{1}{2}q^3$
total frequency in each sex	1	$\frac{1}{2}p^2$	pq	$\frac{1}{2}q^2$	$\frac{1}{2}p$	$\frac{1}{2}q$
		p^2	$2pq$	q^2	p	q

male parents, so that $q_{1f} = \frac{1}{2}(q_{0m} + q_{0f})$. Consequently, if q_{0m} is greater than q_{0f} the new frequency q_{1m} will be less than q_{1f}, and vice versa. The alternation continues until equilibrium is reached. Equilibrium is achieved because every generation of females receives some genes from male parents. The difference $q_{1m} - q_{1f}$ is equal to $q_{0f} - \frac{1}{2}(q_{0m} + q_{0f})$, so that the ratio of the difference between the sexes in succesive generations, or $(q_{1m} - q_{1f})/(q_{0f} - q_{0m})$, is equal to $\frac{1}{2}$. The difference in frequency is therefore halved per generation and in theory the equilibrium is reached after an infinite time (Fig. 2.1).

We can see what the equilibrium value is by considering the values of q_m and q_f, shown in Fig. 2.1, over two generations. Between the vertical coordinates at n_0 and n_1 there are two similar triangles q_{0f}, b, q_{0m} and q_{1f}, b, q_{1m}. Since $q_{0f} - q_{0m}$ is twice $q_{1m} - q_{1f}$ the

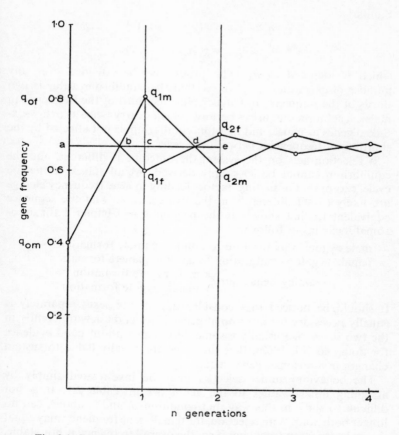

Fig. 2.1. Progress of gene frequency at a sex-linked locus in the absence of selection. The population is started with the frequency q_f of females different from the frequency q_m in males. Over subsequent generations the frequencies in the two sexes converge to an equilibrium state at q_a.

altitude ab of the first triangle is twice the altitude cb of the second. Similarly, cd is twice the length of ed. The point of intersection of the lines from q_{0f} to q_{1f} and q_{0m} to q_{1m} is therefore defined by the equation of the straight line from q_{0m} to q_{1m} at $n = \frac{2}{3}$. It is

$$q_a = q_{0m} + \tfrac{2}{3}(q_{1m} - q_{0m})$$

$$= \tfrac{2}{3}q_{0f} + \tfrac{1}{3}q_{0m}$$

Similarly,

$$q_c = q_{1m} + \tfrac{2}{3}(q_{2m} - q_{1m})$$
$$= q_{0f} + \tfrac{2}{3}(\tfrac{1}{2}q_{0m} - \tfrac{1}{2}q_{0f})$$

which is identical to q_a. The proof may be extended over any number of generations; so we see that the equilibrium value is two thirds of the frequency in females plus one third of the frequency in males, a solution that arises because females carry twice as many sex-linked genes as males, and that this equilibrium is not affected by the shuttling of frequencies between the sexes.

If selection acts on the system the outcome is different, and the equilibrium cannot be so simply derived. At all stages of the life cycle, except that of mating, factors leading to gene frequency change are likely to act differently in the two sexes, so that the sequence equivalent to that shown at the beginning of Chapter 1 for autosomal loci runs as follows:

male zygote → maturation to adult → gamete formation ＼
female zygote → maturation to adult → gamete formation ／ →

→ mating behaviour 〈 male zygote formation
　　　　　　　　　　 female zygote formation

It should be noticed that consideration of the sexes separately is equally necessary for autosomal genes if selection acts differently in the two sexes. We usually assume, although without good evidence for doing so (Li, 1963), that the sexes are equally liable to sustain changes in autosomal gene frequency.

The behaviour under selection may be investigated simply by assigning fitness values to the genotypes in Table 2.1. It is not difficult to show in this way that conditions b and c, above, can no longer both hold. With selection, the female gene frequency may equal the male gene frequency, but if so, the overall frequency is not stable. If the gene frequency is constant then it differs in the two sexes. This kind of notation has been used by Mandel (1959). Bennett (1958), Li (1967) and others have shown that the study of sex-linked situations is facilitated by expressing gene frequencies as ratios of one to another, instead of using the gene frequencies themselves. Haldane and Jayakar (1964) suggest a treatment which provides convenient solutions.

Let the ratio of A_1 to A_2 in eggs in a given generation be $u : 1$, and the ratio in spermatozoa be $v : 1$. If mating is at random the genotype frequencies after one generation are the products of the frequencies in the gametes from which the zygotes have been derived. In the notation of these ratios the *frequencies* of A_1, A_2 and Y bearing gametes in the males are thus $v/2(1 + v)$, $1/2(1 + v)$ and $(1 + v)/2$

$(1 + v)$. The frequencies of A_1 and A_2 gametes in the females are $u/(1 + u)$ and $1/(1 + u)$. We therefore have the following matrix of frequencies of zygotes, in which the common denominator $2(1 + u)$ $(1 + v)$ has been left out throughout.

female gametes		male gametes		
		A_1 $v/2(1 + v)$	A_2 $1/2(1 + v)$	Y $(1 + v)/2(1 + v)$
A_1	$\dfrac{u}{(1 + u)}$	uv	u	$u(1 + v)$
A_2	$\dfrac{1}{(1 + u)}$	v	1	$(1 + v)$
		female progeny		male progeny

The classes of individuals are therefore
A_1A_1, A_1A_2, A_2A_2 in females and A_1, A_2 in males in the ratios
$uv : u + v : 1$ and $u : 1$
Let these classes have fitnesses:
$1 + f$, $1 + h$, $1 - f$ in females   and $1 + m$, $1 - m$ in males
It may then be shown that in the next generation

$$u_1 = \frac{2uv(1 + f) + (u + v)(1 + h)}{(u + v)(1 + h) + 2(1 - f)} \tag{1}$$

and

$$v_1 = \frac{u(1 + m)}{1 - m} \tag{2}$$

At a stable non-trivial equilibrium (one where the frequency is not zero or one) these equations provide

$$\bar{u} = \frac{h - fm + f + m}{h - fm - f - m} \tag{3}$$

and

$$\bar{v} = \frac{\bar{u}(1 + m)}{1 - m} \tag{4}$$

Calling the frequency of A_1 p and the frequency of A_2 q, the equilibrium frequencies of A_2 in eggs ($\bar{q}_e$) and spermatozoa ($\bar{q}_s$) are

$$\bar{q}_e = \frac{1}{2}\left(\frac{h - fm - f - m}{h - fm}\right) \tag{5}$$

and

$$\bar{q}_s = \frac{(1 - m)(h - fm - f - m)}{2(h + m^2)} \tag{6}$$

The condition of equilibrium will only be met by values of h, f and m which satisfy the inequalities

$$(1 + f)(1 + m) < 1 + h > (1 - f)(1 - m) \tag{7}$$

Equations (3), (4) and (7) are identical to Mandel's equations (3), (4) and (5). Equations (1) and (2) are equivalent to (89a) of Li (1967) and equation (3) is the same as his (89b).

We thus have a condition for equilibrium similar to the requirement that heterozygote fitness be greater than the fitness of either homozygote for an autosomal locus; but here each of the fitnesses compared is a product of fitnesses in the two sexes. The notation obscures this correspondence for the heterozygous condition. The complete term here is $(1 + h) \times \frac{1}{2}[(1 + m) + (1 - m)]$, that is, the product of the fitness of the heterozygous female and the mean of the fitnesses of the two male hemizygotes (Li, 1967), which, of course, reduces to $1 + h$.

Study of the requirements will show that the stable state can be reached in two ways. There may be heterozygote advantage in the females ($h > |f|$), in which case the fitness of the male hemizygotes and female homozygotes may vary in the same or in opposite directions. Alternatively, a balance may be maintained and keep a mutant gene in the population if the mutant hemizygote and homozygote differ from the normal in opposite directions. A wider range of fitness values will produce balance when f and m have opposite signs than when they have the same sign.

Haldane and Jayakar refer to the case of the Glucose–6–Phosphate Dehydrogenase polymorphism in man. The G^+ locus is sex-linked, and the mutant G causes low activity of the enzyme. It is fully expressed in the GG homozygote and G hemizygote, partially so in the G^+G heterozygous female. The principal disadvantageous effect is haemolytic disease at birth and on exposure to certain kinds of food and drugs in later life. The gene frequency is usually low, but in parts of the world where *falciparum* malaria is endemic this factor, like the S and C haemoglobin genes, is often common. Frequencies of around 20% occur in Africa, and 12% in the Philippine Islands. Moreover, in Sardinia, where malaria was common in low-lying areas, the frequency rises to 40% on the coast but is only about 4% in the mountainous regions little affected by the insect-borne disease (see

Motulsky, 1963). The Sardinian distribution is closely paralleled by that of the Thalassaemia gene, which is also thought to confer resistance to malaria (Ceppellini, 1955; see also Livingstone, 1967).

Allison (1961) and Motulsky have both discussed the maintenance of the polymorphism, suggesting that although the GG homozygote and G hemizygote are always less fit than typicals, or at least than heterozygotes, the heterozygote has an advantage in malarial regions; in other words, that h is greater than $|f|$ while m and f have much the same values.

At the moment the existence of heterosis in females has not been demonstrated. Heterozygotes produce some red cells of each type, while homozygotes have all of one or all of the other, so that as an environment for malarial sporozoites the heterozygote is intermediate between the other two. By analogy with the sickle cell haemoglobin system, resistance to malaria should be conferred on both heterozygote and the G-bearing homozygote and hemizygote. If the latter genotypes have a net fitness as high as the heterozygote we should expect the gene to go to fixation. If not, there will be polymorphism with heterozygote advantage. A balance without heterosis could arise if, for example, a lesser robustness of males compared with females results in the G hemizygote having a net disadvantage compared with G^+, while the GG homozygote is at an advantage. Alternatively, there could be differential survival of spermatozoa from G^+- and G-bearing males. As a numerical example of the relation between the coefficients, suppose the GG homozygote has an advantage of a little over 10 % compared with G^+G^+ in a malarial area. This figure is not large compared with the difference in fitness found with the sickle cell polymorphism. The value of f is then 0·05. If we suppose the heterozygote to be intermediate ($h = 0$), polymorphism will be maintained if $1·05(1 + m) < 1 > 0·95(1 - m)$; from which we find that m must lie between $-0·048$ and $-0·053$. That is, the advantage to G^+ males must be from 10 to 11 % approximately. If there is heterosis, say with $h = 0·1$, the range of values for m is greatly increased: to between $+0·048$ and $-0·158$. The fitness of G^+ can vary from a 9 % disadvantage to a 37 % advantage and polymorphism will still be maintained.

Finally, it may be noticed that because selection of a sex-linked gene is akin to gametic selection in the hemizygous sex, the response to selection is more rapid than for an autosomal locus. At any gene frequency, only half as many deleterious recessive genes are protected in the heterozygous state as at an autosomal locus, while all such genes must be acted upon in every generation of purely gametic selection. This difference is clearly reflected in the curves that are obtained from graphical analysis of the different systems.

3

EQUILIBRIA AND POLYMORPHISM

The most notable feature of the genetics of natural populations is not the effectiveness with which frequencies are changed by selection and old genes replaced by new, but the prevalence of apparently stable polymorphisms. The term polymorphism implies the coexistence of different forms at the same stage of development in a population (cf. Kennedy, 1961). As a rule we are interested in genetic polymorphism, but as with selective coefficients the arguments are also applicable in certain experimental circumstances to morphs which are not genetically determined. Genetic polymorphism is defined by Ford (1940) as 'the occurrence together in the same locality of two or more discontinuous forms of a species in such proportions that the rarest of them cannot be maintained merely by recurrent mutation'. The term is usually used when discussing different forms which are under the control of alleles of a major gene, but if, for example, we can infer from experiments on quantitative inheritance that a high degree of heterozygosity is maintained, it is reasonable to consider the loci concerned to be polymorphic, even though the character controlled does not segregate into distinct classes.

E. B. Ford's definition excludes mutation as an agency leading to polymorphism. A deleterious mutant will be eliminated rapidly when it appears, so that the frequency of such mutants will never be more than one in a few thousand, while an advantageous mutant should replace the previously typical allele. But from a theoretical point of view it is possible for alleles with almost exactly equal fitness to establish a polymorphism, the equilibrium frequency being determined by forward and reverse mutation rates. Perhaps there are some genes available to act as modifiers of the expression of a major locus which fall into this category. Certainly the substitution of a single base by another in a segment of the DNA of a cistron cannot always lead to a large fitness difference.

The effect of recurrent mutation

The requirement for the continued coexistence of two forms with frequencies p and q in a population is that $\Delta q = 0$ when q lies between zero and one. In the case of mutation, suppose that genes A_1 and A_2 are present at frequencies p and q and that the mutation $A_1 \to A_2$ occurs with frequency u, the back mutation $A_1 \leftarrow A_2$ with frequency v. Then

$$q_1 = (1 - v)q_0 + up$$

so that

$$\Delta q = up - vq \tag{1}$$

Equating (1) to zero we find

$$\hat{q} = \frac{u}{u + v} \tag{2}$$

where $\hat{q}$ (or q hat) means the equilibrium frequency of q. If the forward and back mutation rates are equal, then $q = 0\cdot5$. The rate of approach to equilibrium is found by calculating

$$\frac{d \, \Delta q}{dq} = -(u + v) \tag{3}$$

The reasoning behind this procedure is explained in Chapter 6. Equation (3) indicates that the rate of change in frequency at different frequencies is a constant depending on the overall instability of the locus, and since $d \, \Delta q/dq$ is negative that the equilibrium is stable, (compare equation 15, Chapter 1). As an example, the mutant 'histidineless' (h^-) in E. coli has a mutation rate of $2\cdot9 \times 10^{-8}$ to h^+, while the rate $h^+ \to h^-$ is $1\cdot2 \times 10^{-6}$ (Atwood, Schneider and Ryan, 1951). As is commonly the case, the change to a non-functioning form is more frequent than the reverse change to a functioning condition. This may be expected, since the defective mutant is any one of a possibly large array of configurations at the locus; while the optimal form is just one sequence from the array. The equilibrium frequency of h^+ should therefore be $2\cdot9/123$ or about 2%. No examples of a balance achieved by mutation are available from higher organisms but they might be expected to occur under the rather constant experimental conditions in which bacteria are cultured. In fact, even here selective neutrality is not realized and the equilibrium frequency observed in the experiments of these authors was of the order of 10^{-6}. The variation in frequency of h^- which occurred over a period of time appears to be associated with selection for mutants increasing the rate of increase of numbers in the culture medium.

Gametic selection

When there is selection $d \, \Delta q/dq$ is usually a function of q, and the condition of stability is found by obtaining its slope about $\hat{q}$. Under selection Δq must always be zero at $q = 0$ and $q = 1$, so that the general relation to be expected must be something like Fig. 3.1. General conditions for equilibrium are discussed by Li (1955b, 1967) and Lewontin (1958).

Change in frequency of a pair of morphs, or gametes, under selection may be represented by equations (6) and (19) of Chapter 1, namely

$$\Delta q = \frac{-spq}{1 - sq}$$

from which it is seen that $\hat{q}$ lies between zero and unity if $-spq = 0$. This condition can only occur if s is always zero, a solution that is not of interest, or if s is some varying function of q which has a zero value at $\hat{q}$. Polymorphism can therefore only be maintained with gametic selection if the selection is frequency-dependent. For the equilibrium to be stable s has to be positive for large and negative for small q.

Zygotic selection

Now consider the outcome of selection on zygotes or diploid organisms. The general expression for a locus with two alleles is (Table 1.4)

$$\Delta q = \frac{pq[(s_1 p - s_3 q) - h(1 - 2q)]}{1 - s_1 p^2 - s_3 q^2 - 2hpq} \tag{4}$$

so that for equilibrium $\hat{q} = 1$ or 0, or

$$(s_1 \hat{p} - s_3 \hat{q}) = h(1 - 2\hat{q})$$

and

$$\hat{q} = \frac{h - s_1}{2h - s_1 - s_3} \tag{5}$$

a relation which was derived by Fisher. Since the mean fitness maximizes at the non-trivial equilibrium under these conditions, we can find the same relation by equating $d\overline{w}/dq$ to zero. From (4) we see that $\overline{w} = 1 - s_1 p^2 - s_3 q^2 - 2hpq$ and therefore

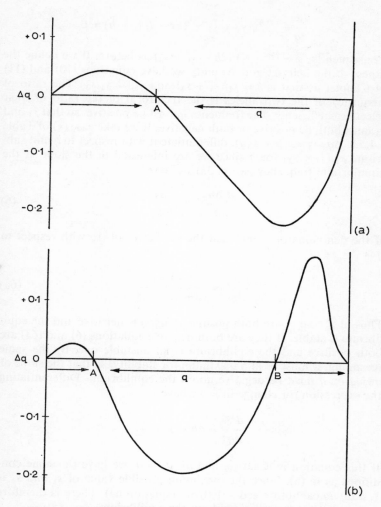

Fig. 3.1. Graphs illustrating types of change in gene frequency under selection. A is a point of stable non-trivial equilibrium, B is an unstable one. The arrows show the direction of change for populations at points between these equilibria. The selection giving rise to these two curves is described in Chapter 5 (p. 111). Both involve selection in different niches, but curve a is characteristic of any situation where the heterozygote has an advantage over both homozygotes.

$$\frac{d\bar{w}}{dq} = -2q(s_1 + s_3 - 2h) + 2(s_1 - h) = 0$$

Consequently, $\hat{q} = (h - s_1)/(2h - s_1 - s_3)$ as before. If we define the fitness of the heterozygote as unity we have equations (10) and (11) of Chapter 1, that is $\hat{q} = s_1/(s_1 + s_3)$ and $\hat{q}/\hat{p} = s_1/s_3$. The ratio of frequency of the two alleles is the reciprocal of the ratio of their selective coefficients. The frequency q must be positive, so that s_1 and s_3 must both be positive or both be negative. If we take case (v) of Table 1.4, i.e. $\Delta q = pq(s_1 p - s_3 q)$, differentiate it with respect to q and substitute $s_1/(s_1 + s_3)$ for q since we are interested in the slope at the equilibriuim frequency (see p. 33) we have

$$\frac{d\,\Delta q}{dq} = \frac{-s_1 s_3}{s_1 + s_3} \tag{6}$$

If the denominator is included the derivative of Δq with respect to q is

$$\frac{d\,\Delta q}{dq} = \frac{-s_1 s_3}{s_1 + s_3 - s_1 s_3} \tag{6a}$$

Thus, if s_1 and s_3 are both positive $d\,\Delta q/dq$ is negative and the equilibrium is stable. If they are both negative equations (6) and (6a) are both positive and the equilibrium is an unstable one. By the same reasoning, $\bar{w}$ must be at a maximum for stability, so that the slope of $d\bar{w}/dq$ on q must be negative about the equilibrium. Differentiating the expression for change in $\bar{w}$ we have

$$\frac{d^2\bar{w}}{dq^2} = 4h - s_1 - s_3$$

If this equation is negative, and $d\bar{w}/dq = 0$, we have the same conditions as in (6). Since the maximum possible value of s_1 and s_3 is 1, $d\,\Delta q/dq$ cannot exceed -1 (from equation 6a). There is therefore no possibility of oscillation about the equilibrium.

Under zygotic selection, then, a stable equilibrium for two alleles can be attained with a single constant selection pressure acting on the genotypes, provided the heterozygote is fitter than either homozygote. Nevertheless, equation (7) of Chapter 1 shows that, just as with gametic selection, the selective value to be attached to the *allele* is still frequency-dependent, because the frequency of genotypes is a function of gene frequency. In Fig. 3.2 the average fitnesses of the alleles, which are $(w_1 p^2 + pq)$ and $(w_3 q^2 + pq)$ respectively, are

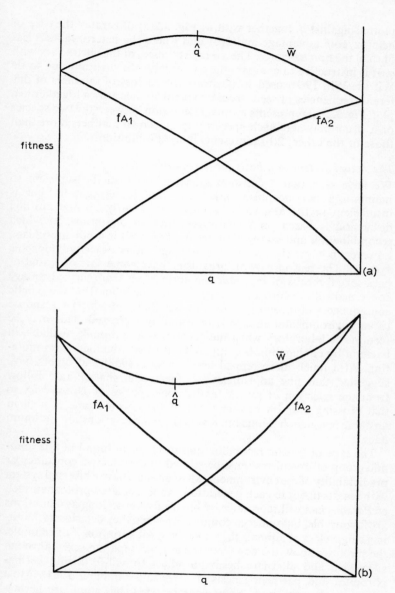

Fig. 3.2. Curves for fitness of alleles when there is heterozygote advantage (a) and disadvantage (b). Allele A_1 has frequency p and A_2 frequency q. $fA_1 = w_1 p^2 + pq$. $fA_2 = w_3 q^2 + pq$ and $\bar{w} = fA_1 + fA_2$. Equilibrium occurs where $d\bar{w}/dq = 0$.

plotted against q, together with $\bar{w}$. Fig. 3.2(a) illustrates the case of heterozygote advantage and 2(b) that where the heterozygote is less fit than the homozygotes. The average fitnesses of the alleles vary with q. It is instructive to compare the curves with the diagram due to de Wit (de Wit, 1960) used to illustrate the success of mixtures of different frequencies of seeds from two plant species grown together in a plot. Fig. 3.2 (a) represents mutual facilitation between the two species or a situation where each species affects its own numbers more than those of the other; 2(b) represents mutual inhibition.

The sexual system as a basis for polymorphism

We have seen that a constant selective force can be sufficient to maintain a polymorphism provided there is segregation, but is insufficient in the absence of segregation. Nearly all the ensuing discussion is based on sexually reproducing organisms, in which recombination and segregation occurs. Sex itself is a polymorphism, however, not merely the vehicle that maintains other polymorphic systems. The reason why all organisms have some sort of sexual or para-sexual means of reproduction springs not from any advantage concerned with the process of reproduction itself but from the genetic consequences of recombination. The possibility of adaptive response to an environmental change is enormously increased. This may be seen in the laboratory when the genetic variation made available in bacteria by conjugation is compared with that obtained from mutation. At a much more refined level Carson (1958) and others have suggested that fine adjustments to the recombination rate follow from the selection of inversions in some species of *Drosophila*, so that crossing-over on certain parts of the chromosome is high in unstable marginal habitats but low in the relatively constant optimum ones.

The type of genetic recombination mechanism found in a taxonomic group of organisms must be related to the effective constancy or predictability of the environment. Most animals have a bisexual system with reassortment in each generation, while a ciliated protozoan such as *Paramecium* will reproduce by binary fission so long as conditions are favourable, but undergo conjugation when they deteriorate. Within a single class of animals there may be great variation. For example, the molluscs show the complete range from separate sexes, through successive and alternate hermaphrodites to self-fertilizing hermaphrodites and parthenogenesis (Fretter and Graham, 1964). Most prosobranch gastropods have separate sexes, but some are hermaphrodite, the species involved coming from several genera and usually having habits of life such as parasitism or restricted mobility. Mate finding is therefore more difficult for them than for their bisexual relatives. The slipper limpet *Crepidula fornicata* is an example of a protandrous hermaphrodite species, in which clusters of individuals live settled upon one another. There is apparently a substance secreted by the mature females at the bottom of the pile which maintains the maleness of younger individuals at the top, so that environmental regulation of the functioning sex of an individual occurs throughout life. Usually, however, the pattern, once established, continues until death.

Many hermaphrodites are self-fertile but some are obligate cross-fertilizers. Among terrestrial gastropods some slugs such as the large black *Arion ater* may on occasion self-fertilize, while others, such as *Agriolimax reticulatus*, are autosterile. In the helicid snails autosterility is the rule. *Arianta arbustorum* has eggs and sperm developing side by side in the early part of the season, separated by a layer of germinal epithelium that covers the eggs while sperm are free to be released. Later, the remaining male gametes degenerate, the separating layer breaks down, and the eggs can be fertilized. Copulation between two individuals is reciprocal. With this pattern of development sperm must survive storage for long periods of time to allow fertilization to take place. As a consequence, sperm from different copulations may get mixed before fertilizing a batch of eggs, so that from a genetic point of view the effective population size is increased. Murray (1964) has estimated that in *Cepaea nemoralis* the average number of matings contributing sperm to each brood is at least two. Both hermaphroditism and the sperm mixing are understandable in species that are often dispersed in small isolated colonies. In surveying the sexual mechanisms displayed by molluscs, it appears that where the ecology of a species demands it, because of the low probability of finding a mate, the system is correspondingly modified.

Similarly, most crustacea are bisexual with a genetic switch mechanism, but there are exceptions among sessile forms including parasites, and in species which require a resistant overwintering stage. In the Cladocera seasonal change occurs from sexual reproduction to parthenogenesis. The barnacles and their parasitic relatives the *Rhizocephala* are hermaphorodite (with a few exceptions) while some parasitic isopods have a facultative sex-determining mechanism that depends on the presence or absence of another parasite already established on the host when a new parasite arrives. If the host is unparasitized the settling individual develops into a female, but if a female is already present the second parasite becomes a male. The consequence of these three crustacean sexual systems can be glimpsed if we consider their relative success under simple conditions of parasitization. Suppose that a host species is attacked at random, and that the density and host-finding ability of the parasite species

c

is such that the mean frequency of parasitization is m per host. Reproduction takes place on the host. The numbers of parasites per host will follow the Poisson distribution, so that we can measure the success of the parasites by estimating the probability of their being in a position to leave offspring. If the parasite is a self-fertilizing hermaphrodite, or is parthenogenetic, then this condition is equivalent to the probability of finding a host, which is $1 - e^{-m}$. If the parasite is a facultative bisexual one success is represented by the occurrence of at least two individuals on a host, which has the probability of occurrence of $1 - (1 + m)e^{-m}$. If the parasite is a conventional bisexual one, on the other hand, it can only reproduce when unlike sexes are present together, and the probability of successful combinations is $\sum_{n=2}^{n=\infty} (1 - 2^{1-n})(m^n/n!e^m)$. The relative success of the three systems varies with m. If the mean frequency of parasitization is 0·1, for example, the hermaphrodite is forty times better off than the genetically bisexual species, while the facultative species is almost twice as successful. If $m = 1$, however, the respective factors are 4 and 1·7. Their values continue to converge as m increases.

I am grateful to Dr D. J. Crisp for bringing to my attention a striking instance of the effect the breeding system can have on success. The two barnacles *Balanus amphitrite* and *Elminius modestus* are recent introductions to the British coast. *Balanus amphitrite* will not tolerate the cold coastal water here: it is transported in the cooling systems of steamships and lives where there are warm water effluents from power stations and similar intallations. Nevertheless, because it is a self-fertile hermaphrodite it is not uncommonly established in such artificial warm environments. *Elminius modestus* is an Australian species that is a very successful colonist spreading along the British and European coastlines. It is self-sterile and did not appear in the region until about 1940 (Crisp, 1958), arriving then probably because the outbreak of the war led to the congregation at one time in British ports of unusually large numbers of ships from the southern hemisphere. Only when the density of adults was raised in this way did the species achieve the necessary critical level for successful establishment, though ships carrying barnacles from Australia must have been arriving for about two hundred years.

The problem of breeding system and rarity is discussed for free-living species by Andrewartha and Birch (1954), Mosimann (1958) and others, the conclusions being essentially similar. The uncommonness of hermaphrodites and asexual organisms suggests that the advantages of these modes of life are more than offset by the drawbacks associated with a more complex structural organization, when it occurs, and with reduction of genetic recombination. Parasites combine a low probability of meeting a mate with a physical environment which is more than usually stable, even though there may be defence mechanisms in the host to contend with. For them the relative constancy of the environment reduces the need for genetic recombination, but for most animals genetic reassortment appears to be the most important requirement.

Provided the average fitness of an organism which can undergo recombination is greater than the fitness of one which cannot there will be selection for outbreeding mechanisms. Such mechanisms can be seen as having two components, a positive one making exchange possible and one that inhibits selfing or mating with a member of the same clone. In bacteria the existence of transformation shows that free bacterial DNA has physical or chemical properties enabling it to be introduced into a living individual at specific sites on the cell surface and to be incorporated into the chromosome. The episome, free or attached to the chromosome, fulfils the second requirement in conjugation. These two aspects do not by themselves explain why there should be no more than two sexes as a rule, and not a series of self-incompatible forms all capable of genetic donation or exchange with unlike individuals. There are examples approaching this situation, such as tristyly and other breeding systems in plants and the mating types of *Paramecium*, which may be numerous within a variety in *P. bursaria* although there are only two per variety in *P. aurelia* (Beale, 1954). The *s* allele systems in plants such as *Nicotiana* and *Oenothera* also have the same effect. They are secondary developments based on the primary sexual system, but where the plant is potentially a self-fertile hermaphrodite the result is functionally similar to the multiple mating types recorded for *Paramecium bursaria*. Systems in which there are several self-incompatible but cross-fertile types are not common, however. One reason might be connected with the fact that the majority of animals and the higher plants are diploid. A primary self-sterile, cross-fertile system preventing fusion of gametes from the same individual would be difficult to effect at the haploid level by a simple genetic switch. Whenever an organism was heterozygous for the genes which inhibit fusion it would produce gametes that could unite. A solution to this problem is available if the genetic mechanism operates at the diploid level to determine the kind of gamete which is produced, or its time of development, so that gametes coming from an individual of one genotype only unite with those from another diploid genotype. The differentiation of gametes into male and female under the control of a genetic locus acting in the diploid stage provides such a mechanism. The mechanism entails the possibility of at least three genotypes so that there could in theory be three sexes rather than two. When the switch depends on one gametic type being motile and the other sessile, a difference which is made more marked if there is also food storage in the egg, there can be only two fully efficient morphological types. It would be possible for one sex, for example the female, to be homozygous for either A_1A_1 or A_2A_2 while the males are A_1A_2, but once the sexual system is established the two forces serving to maintain it are the increase in mean fitness resulting from out-breeding (cf. Fisher, 1930; Crow and Kimura, 1965; Maynard Smith, 1968b) and the tendency for the sex ratio to equalize (for the latter see Fisher, 1930; Shaw and Mohler, 1952; Edwards, 1962). A system with two female types will not produce more recombinants than one where there is only one, and the sex ratio remains at 1 : 1 whatever the ratio of A_1A_1 to A_2A_2, so that if a fitness difference between the two kinds of female arises from another quarter, one of them will be lost.

If the distinction between the sexes depends on the dominant effect of a gene, the cross of the heterozygote to the recessive provides a 1 : 1 sex ratio, while the cross involving the dominant homozygote gives rise to unisexual progeny. Selection for a 1 : 1 ratio will there-fore favour any genes lowering the fitness of the dominant homo-zygote.

The self-sterility systems in plants do represent a gene-controlled multiple outbreeding system. Development or failure of the pollen tube depends on the genotype of the diploid style. Only heterozygotes can exist because no seed is fertilized by pollen carrying a gene present in the style. The minimum number of alleles required is therefore three. The origin of the system appears to require the existence of sterile homozygotes reproducing vegetatively for long periods of time until an opportunity arises to cross with a plant of a different genotype. A multiple self-sterility system is advantageous if the chance of encounter with gametes from another individual is small, because the probability that those encountered are of an unlike type is increased. Similar systems may be discovered in animals when the genetics of sessile groups such as coelenterates, which go in for asexual reproduction, is studied.

The point to be emphasized in relation to the present discussion is that although the XY sex-determining system of most animals is a genetic polymorphism which has developed as a result of natural selection it is one that is highly evolved and very stable, and for this reason cannot obviously be seen to have arisen from the action of the same kind of mechanisms as other polymorphisms. Furthermore, being so well established it plays a necessary part in the maintenance of most of the others. Evidence of the kind of forces that lead to the development of sex is to be found in plants but hardly at all in animals. Among plants one of the well-known studies is that of *Primula vulgaris* by Crosby and others. In this species a pin, a thrum and a homostyle form of flower can occur, the latter resulting from crossing over within the super-gene controlling flower type (see the account in Ford, 1964). The structure of the flowers is such as to give the homostyle an advantage and the thrum form a disadvantage in pollination, so that the homostyle should increase in frequency even in the face of quite large reduction of viability or fitness associated with selfing (Crosby, 1949, gives a derivation of the algebra involved, and the example is discussed in terms of stochastic simulation by Bodmer, 1960). That the homostyle is not at a high frequency everywhere shows that the position is more complicated, and indicates that there are further factors reducing the fitness. There are only a few examples in animals, such as the poecilid fishes and the moth *Lymantria dispar*, where the control of sex is so variable that we can get any idea of the kind of balance of fitnesses leading to the development, and in the long run serving to maintain, the sex supergene polymorphism.

Types of polymorphism

We may now consider in more detail the equilibrium states found with various kinds of polymorphism. Comparatively little attention has been given to the systematic study of the range of systems that may be involved, but the question is discussed by Williamson (1958), Haldane and Jayakar (1963), Li (1967) and others. The existence of an equilibrium gene frequency implies a balance between forces of some kind tending to increase the commonness of one of a pair of alleles and forces tending to reduce it. We may therefore look for situations where there is a conflict between two or more forces affecting the locus. The most obvious sources are ones where different selective agents act in opposite directions, or where the effect of selection is counteracted by that of mutation. In the case of heterozygote advantage, which we often think of first as the cause of polymorphism, the source of the conflict is in fact less obvious. It arises from the interaction of selection favouring a heterozygous genotype and the effect of Mendelian segregation during sexual reproduction, which repeatedly throws up anew the relatively unfit homozygotes. When polymorphism results from a single selective force, the direction of action of which depends on morph frequency, then that frequency is one of the pair of factors involved. The frequency maintains the balance by acting as a negative feedback, but it no longer accords with everyday practice to speak of frequency as one of the antagonists in a contest.

The separation of frequency-independent from frequency-dependent forces is the most important distinction to be made with respect to selection. If the polymorphism is such that a selective value ascribed to a genotype is constant, or varies independently of gene and genotype frequency, then it must be opposed by another selective force or by some property of the system such as segregation. If it is a function of gene or genotype frequency then it may contrive a balance unaided. Nevertheless, for the purpose of reviewing the kinds of polymorphism it is convenient to slice the cake in a different way. Every system mentioned so far relies on a change in frequency occurring somewhere in the life cycle other than at mating. A balance of selective forces may be achieved between selection acting entirely at the diploid level or between the haploid and the diploid stages. Frequency-dependent selection may affect gametes or zygotes. Heterosis depends on the segregation of genes when new zygotes are formed following mating, but none of these systems entails non-random mating, and we may classify them together as occurring within a panmictic group.

A population is said to be panmictic if each adult has an equal chance initially of mating with all others. This condition will occur in a closed population if mating is at random and the individuals can diffuse freely throughout the area occupied. It follows that the gene frequency is the same in subsamples taken from any part of the area. In nature panmixia must be rare. It cannot occur in a very large population, even when there is random mating behaviour, if individuals from some of the localities occupied are partially isolated from others by distance. Even in small populations some kind of niche restriction or temporal separation of the breeding stages may limit the freedom of mating between one class of individuals and another. When there is neither geographical nor ecological separation, behavioural factors may operate to restrict the freedom of choice of mates. Panmixia does not occur, therefore, when mating behaviour is non-random, when there is niche specialization or when the population studied receives a recurrent supply of immigrants from beyond its boundaries. These three very common circumstances may all generate polymorphisms, whereas the polymorphisms so far mentioned could develop within a panmictic unit. It will therefore be convenient to consider the two categories separately.

4

POLYMORPHISM WITH PANMIXIA IN A
HOMOGENEOUS ENVIRONMENT

SINGLE CONSTANT SELECTIVE FORCES

The simplest genetic polymorphism in diploids is one maintained by heterozygote advantage. A single selective pressure is involved and the balance follows from elimination of homozygous segregants. The features of this system have already been discussed in the previous chapter, and will only briefly be summarized at this point. When the three genotypes have initial frequencies p^2, $2pq$ and q^2 with fitnesses

$(1 - s_1)$, 1 and $(1 - s_3)$ then $\Delta q = \dfrac{pq(s_1 p - s_3 q)}{1 - s_1 p^2 - s_3 q^2}$. This equation is a

cubic one in q: for small values of s we have $\Delta q = q^3(s_1 + s_3) - q^2(2s_1 + s_3) + s_1 q$. When $\Delta q = 0$ there are three roots, one of them being $q = 0$. The others can be found by dividing each term by q, so that the equation is a quadratic with roots $q = 1$ and $q = s_1/(s_1 + s_3)$. There can be only one non-trivial equilibrium, which is stable when s_1 and s_3 are both positive, unstable when there is heterozygote disadvantage and both coefficients are negative. If $\hat{q}$ is known the ratio of s_1 to s_3 can be found. Conversely, the equilibrium may be determined if the selective coefficients are known. Note, however, that the values of s_1 and s_3 are not obtainable from the equilibrium frequency alone. The minimum amount of information required for this purpose is the gene frequency in three successive generations. For practical purposes the Δq values obtained should be large so that the confidence limits of an estimation are reasonably small. We can then obtain two Δq equations which may be solved for s_1 and s_3.

The approach to equilibrium is one of alternate selection of zygotes followed by random mating which returns the population to the Hardy–Weinberg equilibrium at a new frequency. On the de Finetti diagram the progress of a typical sequence of selection and random mating is represented as in Fig. 4.1. The two curves provide the Hardy–

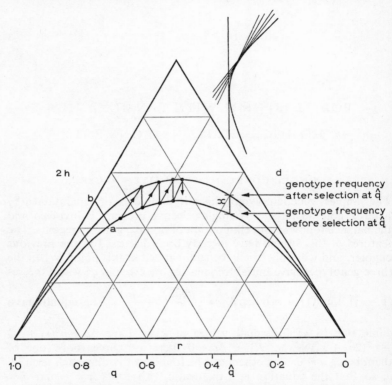

Fig. 4.1. Progress of gene and genotype frequencies under selection. The lower curve represents Hardy–Weinberg frequencies. The upper one is the post-selection curve. Change in frequency proceeds in a stepwise manner as each round of random mating succeeds a round of selection. The equilibrium gene frequency lies on the tangent to the external curve perpendicular to the base line. Modified from Cannings and Edwards (1968). The line ab is explained in Chapter 5, p. 127.

Weinberg frequencies and the zygotic selection frequencies (compare Fig. 1.2). Projections of the lines representing change from gametes to post-selection adults form tangents to curves which have characteristic shapes and positions depending on the selective coefficients (Cannings and Edwards, 1968). The equilibrium, if one exists, lies on the tangent perpendicular to the base of the triangle. The algebraic calculation of absolute values of s_1 and s_3 is represented graphically by finding the appropriate curve to which the projections are tangents, drawing the perpendicular, which provides the ratio $q : p$ and therefore $s_1 : s_3$, and measuring the vertical distance x between the two curves of genotype frequency. This distance is a measure of the magnitudes of the coefficients. From the geometry of the diagram we find

$$s_1 = \frac{x}{p(2pq + x)}, \quad s_2 = \frac{x}{q(2pq + x)}$$

where x is the increment in the frequency of heterozygotes between the offspring of random mating and the post-selection adult stage, and q is the equilibrium frequency.

A polymorphism so maintained in a large population will persist so long as the selective coefficients remain constant. All polymorphisms are in the long run temporary phenomena because species must constantly adapt to a changing environment, but polymorphisms arising from heterozygote advantage may be relatively long lived and widespread since there is only one selective agent involved. If selection favours the heterozygote because, for some reason, an intermediate expression has the highest fitness, then selection acting to modify the expression of one homozygote towards the optimum may lead to loss of the polymorphism even under constant conditions. A change of this kind is probably unlikely to occur because the polymorphism does not in itself add any feature which makes the modification of homozygous expression easier. The most permanent system occurs when the feature favoured is an attribute of the heterozygote alone, such as a hybrid molecule produced by cooperation between both alleles. Evidence for molecular hybrids is not common, but examples are known from man and some animals and plants.

HETEROZYGOTE ADVANTAGE RESULTING FROM A BALANCE OF CONSTANT FORCES

In practice it is usually impossible to distinguish true heterozygote advantage, in which some attribute of the heterozygote is favoured, from a disadvantage to both homozygotes arising from different causes. The distinction may be no more than a semantic one, because even if the heterozygote produces a specific favoured molecule the homozygotes are unlikely to have equal fitnesses, and this fitness difference contributes to determine the position of balance and the size of the coefficients. Once one system is established it is possible for the other to develop. Nevertheless, at the other extreme from the canalized genetic system of true heterosis, a polymorphism could be maintained if one of the homozygotes has an advantage during one stage of the life cycle while the other homozygote is favoured at another.

Suppose that the fitnesses are as follows:

Fitness of three
genotypes

first period of selection	$1 - s_1$	1	1	(1)
second period of selection	1	1	$1 - s_2$	

If the fitness differences are small the overall coefficient for each genotype is almost equal to the sum of the coefficients for each period of selection, and the total change may be calculated using these sums in a simple Δq equation. Selective elimination over the whole period is therefore derived from the fitnesses

$$1 - s_1 \quad 1 \quad 1 - s_2$$

and leads to a stable equilibrium if s_1 and s_2 are both positive (compare equation 13, Chapter 1). Cyclic selection in opposite directions acting on one gene cannot lead to polymorphism if the fitnesses are constant and have the same dominance relation at both stages, but it may do so if the dominance differs between stages, for example if selection acts on different attributes from a pleiotropic array. Thus, the fitnesses

$$1 - s \quad 1 \quad 1$$
$$1 - s' \quad 1 \quad 1$$

cannot give rise to stable equilibrium even if the coefficients have opposite signs, whereas the fitnesses

$$1 - s \quad 1 \quad 1$$
$$1 - s' \quad 1 - s' \quad 1$$

provide over a complete generation

$$1 - s - s' \quad 1 - s' \quad 1$$

which is stable if s' is negative and s positive and greater than zero. Systems of this kind will be affected by fluctuations in the coefficients due to such things as variation in the length of the period over which a certain kind of selection acts, so that the risk of fixation is greater than with true heterozygote advantage.

BALANCE OF GAMETIC AND ZYGOTIC SELECTION

Several examples are known where selective elimination or unequal formation of gametes occurs, to be followed by zygotic selection which leads to a balance. The pattern is not unlike selection on sex-linked genes, where a process similar to gametic selection occurs in one sex while zygotic selection operates in the other.

In the general sense of the term, gametic selection takes place between two kinds of gametes from the total pool. In an animal, such as a species of sea urchin, where eggs and sperm are liberated into the water to undergo a period of free existence before union, a fitness difference between alleles could operate according to the simple rules of gametic selection. Thus, if one allele had 50% fitness relative to the other, selection would operate in the same way whether the parental population was in random mating genotypic proportions or was entirely composed of homozygotes of the two kinds, or had any other composition. Examples observed in nature, on the other hand, are ones where selection takes place only when gametes of the two allelic types are present together in a heterozygote, since the differential elimination operates within the parental individual and involves some sort of competition or antagonism between the types. No change in gene frequency can come about in these circumstances, whether due to differential survival or unequal segregation, if all the adult individuals are homozygous for one pair or the other. Distortion of segregation is usually associated with the process of meiosis, and for this reason is known as meiotic drive.

The progress of gene frequency under the simplest conditions runs as follows. Let the frequencies of two alleles in one generation be p and q. Random mating leads to production of gametes which in the absence of other influences would also be in the ratio p to q. Either these frequencies are not attained because of unequal production or there is differential mortality, so that at the time of zygote formation the frequencies have reached the new ratio p to xq. The factor x is the measure of relative gametic fitness. Upon union the next generation begins with the genotype proportions $p^2 : 2xpq : x^2q^2$, and the environment may influence the genotypes differently, conferring fitnesses w_1, w_2 and w_3.

Suppose that $w_3 = w_2 = 1$. Then

$$q_1 = \frac{x^2q^2 + xpq}{w_1p^2 + 2xpq + x^2q^2} \tag{2}$$

If we write $s_1 = 1 - w_1$, and $g = 1 - x$, then at equilibrium

$$\Delta q = pq[s_1p - g(1 - gq)] = 0$$

whence

$$\hat{q} = \frac{g - s_1}{g^2 - s_1}$$

so that s_1 must be greater than g for equilibrium, and both coefficients require the same sign. When $w_2 = 1$ but w_3 has some other value the equilibrium is

$$\hat{q} = \frac{g - s_1}{g^2 - s_1 - s_3(1 - g)^2} \tag{3}$$

Heterozygote advantage at the zygote stage is not necessary for the balance to be maintained, but it will increase the stability of the locus should fluctuation in the gametic fitness occur.

This example outlines the main features of balance between gametic and zygotic selection. It is hypothetical and relatively unlikely to be discovered in animals, because male and female gametes are so different in structure. A genetically controlled fitness difference in spermatozoa will almost certainly have a different value in ova, and vice versa, so that x varies between the sexes of the parents. An example in which the controlling locus is autosomal and there is a sex difference in the gametic effect is the *tailless* (T) locus of the house mouse (Dunn, 1957; Lewontin and Dunn, 1960). Wild populations are polymorphic for a number of t alleles (distinguished by slightly different effects) which are recessive lethals, $t+$ having a fitness almost identical with $++$ and tt causing death at an early stage of development. Such extreme zygotic selection would quickly lead to elimination of t were it not for the fact that about 95% of the spermatozoa of heterozygous males carry t, and only 5% carry the typical $+$ allele. In the females the segregation ratio is unaffected. The increased likelihood of fertilization by t sperm from heterozygous males has an effect counteracting the elimination of tt homozygotes. There is also some evidence of a greater fitness of heterozygous males over homozygous males.

A theoretical treatment of the t system was provided by Bruck (1956). Let the frequency of t in the adult population before a round of mating be q_0 and let m be the fraction of spermatozoa from $t+$ males which carry t. The genotype frequencies will be defined as d, $2h$ and r. Since tt is lethal the frequency of heterozygotes is $2h_0 = 2q_0$, that of homozygous typicals being $d_0 = 1 - 2q_0$. Mating takes place at random, leading to production of zygotes of three genotypes as follows:

parental mating	frequency	progeny		
		x	$+t$	tt
$+t \times +t$	$4h_0^2$	$2h_0^2(1-m)$	$2h_0^2$	$2h_0^2 m$
$+t \times ++$	$2d_0 h_0$	$d_0 h_0$	—	—
$++ \times +t$	$2d_0 h_0$	$d_0 h_0(1-m)$	$d_0 h_0 m$	—
$++ \times ++$	d_0^2	d_0^2	—	—
Total	1	d_1	$2h_1$	r_1

Using the notation of the previous example and remembering that $w_1 = 1$ and $w_3 = 0$, we may now write for equation (2),

$$q_1 = \frac{xq_0}{p_0 + 2xq_0} = \frac{h_1}{d_1 + 2h_1}$$

This gives a value of $x = \dfrac{1 + 2m(1 - 2q)}{2 - 4mq}$, or $g = \dfrac{1 - 2m}{2 - 4mq}$, showing

that the selection acting on the gametes depends on both m and q. Setting $s_1 = 0$ and $s_3 = 1$ the equilibrium equation (3) is

$$\hat{q} = \frac{g}{2g - 1}$$

Substituting $(1 - 2m)/(2 - 4mq)$ for g we get at the equilibrium

$$4mq^2 - 4mq - (1 - 2m) = 0$$

or

$$q = \frac{1}{2} \pm \frac{\sqrt{[m(1 - m)]}}{2m} \tag{4}$$

Bruck points out that the required solution is $\frac{1}{2} - \dfrac{\sqrt{[m(1 - m)]}}{2m}$ since $q = \frac{1}{2}$ is the highest possible frequency, occurring when all individuals carrying t are heterozygous. There will be a stable equilibrium point between 0 and 0·5 when m lies between 0·5 and 1. Although q rises steadily from 0 to 0·5 as m increases, there is a maximum point for the frequency of heterozygotes at a value of m of about 0·96, above which the heterozygote frequency again falls. This maximum value corresponds to the average value of m found by testing wild mice carrying the various t alleles.

The frequencies of t found by sampling wild populations are usually lower than those expected on the basis of the model. Lewontin and Dunn point out that the reason may be related to the sampling drift effects associated with small effective population size. They suggest that the interaction of selection and random drift promotes a stationary distribution of gene frequency with modes at the two extremes. Most populations at any given time will be homozygous + while some have high frequencies of *t*, but a migrant heterozygous male has a good chance of introducing the gene into a previously monomorphic population, so that the gene spreads from place to place with a relatively high probability of extinction per unit time at each place. The dynamics of the process depends on the ecology and population structure of the mice in a particular area. In order to study the population genetics in nature, mice heterozygous for a *t* allele were introduced into a population on an island where the gene was previously unknown (Anderson, Dunn and Beasley, 1964). Sampling between two and five years later showed the gene to be established and spreading, thus supporting the prediction that the segregational advantage would outweigh the lethal effect, although the rate of spread has been slow.

A similar kind of distorter system occurs in *Drosophila melanogaster* (Hiraizumi, Sandler and Crow, 1960). Known as segregation distorter (SD), the locus involved is autosomal and transmitted in excess by heterozygous males, segregation in female heterozygotes being normal. Usually, but not always, the chromosome bearing SD also carries a recessive lethal gene, so that a polymorphism may be maintained in nature in the same way as the *t* gene in the mouse. Hiraizumi, Sandler and Crow give a detailed discussion of the theory of the system using a more general model than Bruck.

When the segregation distorting gene is sex-linked on the X or Y chromosome, or when an autosomal gene affects the production or viability of X- or Y-bearing sperm, an additional factor enters. The balance of the sexes is disturbed in one direction or the other, so that selection on the sex ratio interacts with the distorter. The outcome may vary depending on a wide range of ecological and genetical properties of the species concerned. At one extreme, introduction of a sex ratio distorter may lead to extinction of a population. A heterogametic male carrying an X-linked distorter causing production of entirely X-bearing sperm will give rise only to daughters. Each will be at least heterozygous for the distorter. The frequency of distorter bearing females is therefore raised, which increases the probability of 100% female broods in future generations. The progression continues until a generation composed entirely of females is produced. Nevertheless, sex ratio distorter polymorphisms are known. Wallace (1948) showed that a balance could be maintained in populations of *D. pseudoobscura* for a gene of this kind, due to interaction similar to that found for t in the mouse and SD in *D. melanogaster*. The 'sex ratio' (SR) factor causes production of nearly entirely female broods by heterozygous males, but under certain environmental conditions homozygous SR females have so low a fitness that the gene is maintained in a polymorphic state. The theory of balance in these conditions is also discussed by Shaw (1959) and Shaw and Mohler (1953).

Distortion of sex ratio unchecked by balancing fitness effects is obviously deleterious. The situation is worse if the distorter is Y-linked and causes an excess of the heterogametic sex to be produced, since no balancing is possible and we have a situation of gametic selection with an advantage to one of the two kinds of Y-bearing gamete. If an autosomal meiotic driving factor has a lower fitness than the typical gene, its introduction to a population may well be harmful. Selection for linked compensating genes tending to establish polymorphism is likely to occur, and selection for suppressors of the driving effect. The t alleles, which are almost always lethal when homozygous, may represent a highly evolved example of balance. The SD factors are associated in the majority of cases with the presence of a recessive lethal within a chromosome inversion, showing how the system can develop. A number of examples are known of meiotic drive suppressor genes which nullify the effect on segregation (e.g. Malagolowkin and Carvalho, 1961; Hickey and Craig, 1966), and indeed Hamilton (1967) has suggested that the genetic inertness or complete absence of the Y-chromosome in so many species may be a consequence of the catastrophic effect, when they occur, of Y-linked distorters.

VARIABLE SELECTIVE FORCE

(a) *Independent of gene or morph frequency*

The argument set out on pp. 73–5 assumes constant selection pressures. If there is fluctuation in selection, however, some situations which would otherwise be unstable may give rise to polymorphism. For example, where $(1 - s)$ and $(1 - s')$ are fitnesses of one recessive homozygote when acted upon by different selection pressures, we could consider these terms to be values with mean coefficients of opposite sign, acting successively during a seasonal cycle. When s and s' are not equal they may give rise to equilibrium if they fluctuate, although they cannot do so when they are constant. If they refer to gametic selection the fact of fluctuation makes little difference because a balance has to be obtained of so exact a nature, the mean effect of one force just cancelling the effect of the other, that the two forms are most unlikely to coexist. The problem is similar to that of the control of numbers by density independent factors alone, which is discussed in Chapter 6. Haldane and Jayakar (1963) have pointed out, however, that when there is segregation of genes the probability of a balance is less remote.

In developing the argument they employ ratios of gene frequencies, as in their discussion of sex-linked selection. Using the terms we have employed so far, let w_i be the fitness of a recessive genotype in generation i, the other genotypes having unit fitness. The gene frequency of the form with variable fitness w is q_i in generation i. A measure of the difference in gene frequency over n generations may then be calculated as

$$\frac{p_n}{q_n} - \frac{p_0}{q_0} = n - \sum_{i=0}^{i=n-1} w_i + \sum_{i=0}^{i=n-1} \frac{w_i q_i (w_i - 1)}{p_i + w_i q_i} \tag{5}$$

A second measure of change in frequency is

$$\log \frac{p_n}{q_n} - \log \frac{p_0}{q_0} = - \sum_{i=0}^{i=n-1} \log w_i - \sum_{i=0}^{i=n-1} \log \left(\frac{p_i + w_i q_i}{w_i} \right) \tag{6}$$

Each equation can conveniently be reduced for values at one extreme of the frequency range. Thus, if p_n and p_0 are both very close to 1 the third term in (5) becomes nearly zero. At this extreme polymorphism required that p_n/q_n be smaller than p_0/q_0, a result achieved if $1 - \frac{1}{n} \sum w_i$ is negative, that is, if the arithmetic mean of the fitness values is greater than 1. The second term of equation (6) likewise diminishes to zero as p becomes very small, so that for a change in frequency near that extreme we have a result favouring polymorphism if $\sum \log w_i$, or $\prod w_i$, is less than 1. Thus, the condition for polymorphism is that the arithmetic mean of successive values of w be greater than unity, while the geometric mean is less than unity.

This condition is restricting, but Haldane and Jayakar point out that it is quite easy to visualize situations where it is met, for example, when a recessive gene has a small advantage in the great majority of generations (or seasons if the animal is annual), but now and again is lethal. In these circumstances the arithmetic mean may readily be greater than 1, while the product of the fitnesses is zero.

The seasonal nature of the environment for the enormous number of animals with annual life cycles may contribute to the maintenance of polymorphisms, or at least slow down the rate at which the morphs are eliminated. This is even more true of bivoltine or multivoltine species, which have the characteristic that the stages of the life cycle in one generation suffer different environmental conditions from those experienced by the equivalent stages of the next generation. When the pair of generations occur in spring and autumn in a temperate region the selective pressures must often tend to act in opposite directions.

Examples where progressive changes in environmental conditions affect successive generations of insects during a year are seen in the classic studies of Dohzhansky and his associates on the inversion polymorphism of *Drosophila pseudoobscura* and in the melanic polymorphism of the ladybird *Adalia bipunctata* (Timofeeff-Ressovsky, 1940; cf. Creed, 1966). Although other factors are certainly involved, a reciprocating selective effect of the seasons may also play a part in maintaining the variation. For a bivoltine species where w_1 is the fitness of a recessive genotype in spring, say, and w_2 is the fitness in the subsequent autumn generation, the condition for polymorphism is that the two values fall within the shaded area on the graph (Fig. 4.2).

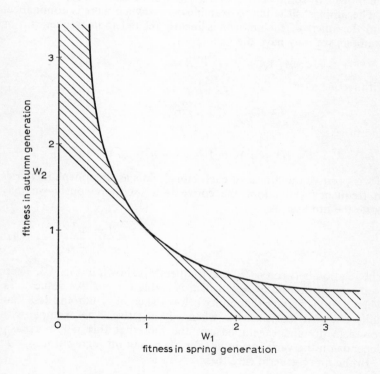

Fig. 4.2. Conditions for polymorphism in a bivoltine species with selection acting on a recessive morph in different directions in the two generations. The successive fitness values are w_1 and w_2. Polymorphism will be maintained if both fitnesses fall within the shaded area. From Haldane and Jayakar (1963).

(b) *Selection varying as a function of gene or morph frequency*

If the selective force varies with frequency, polymorphism need not involve segregation of genotypes. When there is a segregating locus the relative elimination of the segregants may be a function of frequency, and the kind of equilibrium to be expected will be modified. Situations giving rise to frequency-dependent selection have been discussed by Wright and Dobzhansky (1946), Haldane and Jayakar (1963), Clarke (1962), Clarke and O'Donald (1964), Kojima and Yarborough (1967) and others. The most detailed resumé of the general properties of such systems has been made by Li (1967). The basic elements will be discussed here in the context of theoretical models used by Clarke and O'Donald, and Li; then the approach of the models to some natural situations will be considered.

The simplest situation to examine occurs when there is dominance, and the fitness of each morph is linearly related to its frequency. For example, we may have the genotypes

$$A_1A_1 \qquad\qquad A_1A_2 \qquad\qquad A_2A_2$$

with frequencies

$$p_2 \qquad\qquad 2pq \qquad\qquad q^2$$

and fitnesses

$$1 - s(1 - q^2) \qquad 1 - s(1 - q^2) \qquad 1 - sq^2$$

If s is positive the fitness of each morph falls as the morph increases in frequency. To adopt the convention used previously we may write the fitnesses as

$$1 \qquad\qquad\qquad 1 \qquad\qquad \frac{1 - sq^2}{1 - s(1 - q^2)} = wf(q)$$

The relation between $wf(q)$ and frequency is shown in Fig. 4.3. There is an equilibrium at $q^2 = \frac{1}{2}$, which is stable for positive values of s, when $wf(q)$ is greater than unity below the equilibrium and less than unity above it, but unstable when s is negative. The example with complete dominance emphasizes the fact that this polymorphism depends on the variation in $wf(q)$ rather than on segregation.

In the more general case there are fitnesses

$$1 - sp^2 \qquad 1 - 2spq \qquad 1 - sq^2$$

With positive values of s the fitness of the homozygotes falls as their frequencies rise. The heterozygote fitness is likewise lowest at the highest genotype frequency, viz., when $q = 0\cdot5$. When $s = 1$ the rela-

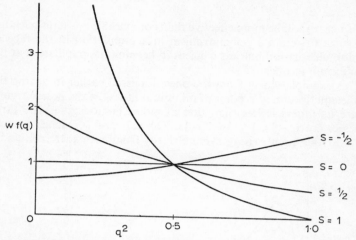

Fig. 4.3. Frequency dependent selective values plotted on q^2 for a system with dominance. The fitness of the recessive phenotype ($wf(q)$) has been made to vary while the fitness of the dominant phenotype is expressed as a constant. $wf(q) = \dfrac{1 - sq^2}{1 - s(1 - q^2)}$.

tion of the selective *coefficients* to q is given by the curves in Fig. 1.1.

As before, the equilibrium is unstable if s is negative, and it is also unstable for large positive values, when fitnesses become negative at some frequencies, which can only mean that the genotype is lethal. The relation of Δq to q is to be found by substituting sp^2, $2spq$ and sq^2 for s_1, h and s_3 in the general equation for constant selection pressures given in Table 1.4, and the equilibrium is most conveniently obtained by substitution in the equilibrium expression to be found there. The result is no longer so easy to derive because there are terms in q of second and third order. From the equation

$$q = \frac{h - s_1}{2h - s_1 - s_3}$$

we obtain by substitution

$$6q^3 - 9q^2 + 5q - 1 = (2q - 1)(3q^2 - 3q + 1) = 0 \qquad (7)$$

The only real root of this equation is $q = \frac{1}{2}$, so that this is the only non-trivial equilibrium. The curve of Δq on q for $s = 1$ is shown in Fig. 4.4, which may be compared with the curve for constant fitnesses in Fig. 3.1(a). Frequency-dependent selection of this kind leads to a more rapid change in frequency near $q = 0$ and $q = 1$, but a relatively shallower approach to the equilibrium. Forces producing such a curve will be more effective than comparable constant selection pressures to retain a polymorphism in a population in the face of random fluctuation but less efficient in keeping the population at the equilibrium frequency.

This example differs from the ones discussed earlier in having the stable equilibrium at a point of minimum fitness of the heterozygote, where its fitness is less than that of either homozygote. The three fitnesses are $1 - \frac{1}{4}s$, $1 - \frac{1}{2}s$, $1 - \frac{1}{4}s$. The value of $\bar{w}$ reaches a maximum at this point, rising symmetrically from each extreme. Under

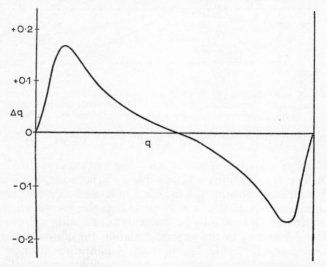

Fig. 4.4. The curve of Δq on q for frequency-dependent selection when the three fitnesses are $(1 - sp^2)$, $(1 - 2spq)$ and $(1 - sq^2)$. The values on the ordinate are for $s = 1$. Compare Fig. 3.1(a).

other models for frequency-dependent selection the heterozygote may have the highest fitness at equilibrium but the change in $\bar{w}$ is no longer the appropriate variable to test for the existence of an equilibrium or its stability. If we write $\bar{w} = \sum fw$, where the f's represent frequencies and the w's the fitnesses, then both parts are functions of q so that the derivative with respect to q is the derivative of a product. It should therefore be written

$$\frac{d\bar{w}}{dq} = \sum w \frac{df}{dq} + \sum f \frac{dw}{dq}$$

Wright (1949) and Li (1955b) show that the behaviour of a frequency dependent polymorphism depends on the term $\sum w \dfrac{df}{dq}$ and not on the sum of the two terms. As stated by Li (1955b) 'the equilibrium value of gene frequency should be such that its effect on the average fitness of the entire population is balanced by its average effect on the individual selective values'.

Under most natural conditions where frequency-dependent selection is operating we may expect frequency-independent selective agents to act as well. The shape of the combined Δq curve will not be simple, and the interaction may lead to interesting new results. For example Clarke and O'Donald showed that a combination of the model described above with heterozygote disadvantage at certain fitness values may produce two stable equilibria with an unstable one between them. Other values for the constant fitnesses lead to a single stable and an unstable equilibrium. If the constant fitness is tending to fix one or other allele the combination may give a curve of Δq on q which is horizontal over a wide range of q but drops to a minimum with a negative value of Δq at one end of the range. The result is to make the allele very similar in fitness for most values of q although heavy selection may still be acting.

Some examples: (a) Inversion frequencies in Drosophila

When comparing the fitnesses of the chromosome arrangements Standard, Arrowhead and Chiricahua in *Drosophila pseudoobscura*, Wright and Dobzhansky (1946) obtained empirical determinations of Δq on q leading to stable equilibria for pairs of karyotypes. Several possible underlying selective regimes are discussed, among them frequency dependent selection. The model employed uses the fitnesses $1 - a + bq$, 1, $1 + a - bq$. Estimations of the constants a and b can be made to provide a good fit to the observed data. The kind of selection which might act in this way in nature would involve niche specialization in a heterogeneous environment. Each genotype may be favoured when it is rare and occupies only its optimum habitat but at a relative disadvantage when it is so common as to overflow into niches favouring the other genotypes. Kojima and Yarborough (1967) also present this hypothesis for frequency-dependent selection. Provided the total population size is constant such selection may strictly be described as frequency-dependent, but if population size fluctuates it is density-dependent. The difference between the two is that a frequency-dependent selective agent acts in the same way at any density, whereas a density-dependent one only operates within the limiting density range. The example is considered more fully in the next chapter, since it depends on environmental heterogeneity.

(b) *Selective predation*

A great many of the adaptations of colour, pattern and form in animals have arisen as a consequence of predation. The predator depends for its existence on its skill in search, discrimination and attack. Its behaviour is continually subject to modification in the light of earlier experience so that there is a feedback element in the response to a choice of prey. Under some conditions this results in a polymorphism.

Several kinds of evidence show that predator response is not linear with change in frequency. Holling (1965) has demonstrated that when an alternative food is available the number of prey individuals eaten by vertebrate predators under experimental conditions varies in a sigmoid fashion with density. The curve is the result of a balance between time and effort put into finding the prey and satiation when high levels of success are achieved. If the total amount of both kinds of food is kept constant then the kind at a low frequency will be under-represented in the sample removed by predation, so that selection of prey is frequency-dependent. The same outcome would occur if the predator has to encounter a certain minimum number of individuals of one kind of prey before they are recognized to be a possible food source. The predator develops a 'searching image' for a particular form when it is common, ignoring it when it is rare. The searching image hypothesis was developed by Tinbergen (1960) and Mook, Mook and Heikens (1960) to account for the behaviour of insectivorous birds when they meet a new form of prey, and several instances of this kind of predation are known. Clarke (1962) applied the argument to morphs within the same species, calling the selection thus applied *apostatic*, because it tends to move the morphs apart from each other in appearance. Selection is most effective when they are unlike and divergence is therefore favoured.

Another relevant kind of predator behaviour is the avoidance of any novel stimulus. Under some circumstances it must be advantageous to a predator to treat rare and unusual objects with circumspection, so that not until a threshold quantity is present are they investigated as a possible source of food. The brain acts as a filtering device, selecting from the stimuli received by the eye those which elicit a reaction. The filter works in a manner that in one way or the other is frequency-dependent. As a rule, events or objects seen very rarely have a low likelihood of leading to active response, activity being channelled to give greatest attention to relatively common events. Optimal response by the predator therefore involves two conflicting behaviour patterns. In some circumstances a reaction of the searching image kind is most suitable, focusing the attention on sources of food most likely to yield an appreciable return, while in others there is a premium on an experimental approach to new food sources which may prove more fruitful than those already exploited. In some species and some circumstances animals should behave in a conservative manner, under-exploiting food objects which are relatively rare, while at other times attention will be focused on rare prey so that it is attacked at least in proportion to its frequency.

Whatever the underlying behavioural cause, a conservative approach will lead to apostatic selection upon a pair of prey species. The selective coefficient of each morph is a function of its frequency and the selection has properties of the same general pattern as those outlined in this section, the outcome in a particular instance depending on the shape of the frequency-dependent function. A number of experimental studies have recently been begun to investigate the effectiveness of apostatic selection in the field. Among the published work Popham (1941) carried out an extensive study of the response of fish to colour forms of an aquatic bug, Allen and Clarke (1968) have studied the reaction of British garden birds to coloured artificial foods, while Mueller (1968) observed the reaction of raptorial birds to different phenotypes of mice. Moment (1962) suggested that this kind of selection accounts for the very wide variety of colour forms found in some polymorphic species, his examples being a brittle star and a small bivalve. If all predators have difficulty finding sufficient similar prey individuals to form a searching image the intensity of predation may be considerably lower than that on a monomorphic prey species. Owen (1963) and Greenwood (1969) have stressed the importance of density, also discussed by Clarke and by Holling, in influencing the nature of the selection exerted by the predators.

The following data illustrate the type of result that can be obtained. They were collected by Miss Pauline Oates and others at Manchester in an undergraduate project on predation of mixtures of red and yellow 'maggots' made of bread, fat and colouring. The predators were birds of city gardens: blackbirds, starlings and house sparrows. In each trial 49 paste models were laid out daily, spaced at intervals of one foot on a 7 × 7 grid. A trial began with four days' presentation of one colour, followed by presentation for four more days of mixtures at frequencies of just over 90%, 70% or 50% of the conditioning colour. The number of individuals removed of each colour was recorded after about half had been eaten. The tests took place in different areas or after the elapse of a long time interval, so that they are independent, two runs for each conditioning colour at each frequency being made.

The results are given in Table 4.1, where the composition of the sample removed by the birds may be compared with the frequencies presented. There is clear evidence of frequency-dependent selection,

showing that the period of conditioning influences the frequency subsequently removed, even though models of both colours must have
been apparent to the predators during the second period.

The four trials with 90% presentation are statistically homo-
geneous, but there is heterogeneity between samples at 70% and at
50%. In ten out of twelve cases a relative excess of the conditioned
colour has been taken, however, and the variability reflects difference
in response between the several groups of predators involved, and no
doubt variation in other uncontrolled aspects of the experiments
such as weather conditions. Analysis of the returns for successive
days suggests that the tendency to take the conditioned colour declines
with time. In nature there is not likely to be a conditioning period
followed by a period of choice, but when a rare morph is at a very
low frequency many individual predators must meet sequences in
this order. A tendency to avoid the rare form even for a short while
will then lead to effective frequency-dependent selection if behaviour
of the kind noticed in this experiment is at all common.

Table 4·1

An experiment on predation of polymorphic prey. Each trial
consists of the display for wild birds of 49 edible coloured
paste models in a 7×7 grid pattern for a total of eight days,
For the first four days all models were red or yellow (the
conditioning period). For the next four days mixtures of red
or yellow in a randomized order were used, at frequencies
of 90, 70 or 50% of the conditioning colour. The table shows
the numbers and frequency taken during the second four-
day period.

	number of conditioning colour removed	total removed	% frequency of conditioning colour removed
red 90%	65 (5)	68	95·6
red, 90%	87 (0)	87	100·0
yellow, 90%	106 (1)	107	99·1
yellow, 90%	44 (0)	44	100·0
red, 70%	107 (9)	116	92·2
red, 70%	79 (29)	108	73·1
yellow, 70%	108 (10)	118	91·5
yellow, 70%	99 (38)	137	72·3
red, 50%	54 (55)	109	49·5
red, 50%	85 (68)	153	55·6
yellow, 50%	93 (59)	152	61·2
yellow, 50%	35 (47)	82	42·7

(c) *Mimicry*

A more complicated kind of frequency-dependent selection is responsible for the development of mimicry. It rests on failure of predators to discriminate between two species when their resemblance is sufficiently close. The outcome also depends on such variables as the extent to which predation affects the numbers of prey in future generations and the way in which discrimination is affected by the availability of alternative food. Excellent accounts of the features of mimetic systems are to be found in Sheppard (1958, 1959), Ford (1964) and Wickler (1968).

The two principal kinds of mimic are the Batesian, an edible form resembling a distasteful model, and the Müllerian, a distasteful member of a group of unpleasant and warningly-coloured species which resemble each other. Since the sensation of distastefulness is relative, varying between species and with hunger level, a species of prey that is a Batesian mimic to one predator may be a Müllerian mimic to another. Nevertheless the characteristic features differ clearly: Müllerian mimicry is the outcome of selection for uniformity and acts upon both the species involved to promote their convergence, while Batesian mimicry favours increasing resemblance of mimic to model. Unlike Müllerian mimicry it may lead to polymorphism. It is with this aspect that we shall deal.

Mimetic polymorphism is the result of selective forces that are frequency-dependent in two respects. They vary with the ratio of mimics to models and with the ratio of mimics to non-mimetic morphs of the edible species. The consequences of these interactions can be displayed by means of a simple theoretical argument. The extent to which we may assess the response to a given ratio depends not only on information about the discriminatory behaviour of the predator, but also on whether the predatory species has a significant influence on the numbers of models and edibles in future generations. If it does not do so, then the ratio of models to the total number in the edible species may be taken, for the sake of the argument, to be constant, so that the ratio of mimics to non-mimetic edibles is a function of the ratio of mimics to models. If predation does affect future numbers, however, no such simple relationship exists. In the discussion which follows it will be assumed that the total number of models and the total number of edibles may be treated as constants, so that only changes in frequency are involved. This would be justified if densities in both species were determined by density-dependent controlling factors acting after predation has taken place. It greatly simplifies the treatment.

As a first step we define M as the frequency of mimetic edibles and $L = 1 - M$ as the frequency of non-mimetic edibles in a polymorphic species. Then the ratio of models to edibles may be defined as $\phi : 1$. If e is the probability of a model being eaten when discovered by a predator, and f is the probability that the predator will eat a non-mimetic edible, the reaction of the predator to a mimic may be expressed in an elementary way by the equation.

$$E = \theta e + (1 - \theta)f$$

where θ is $\phi/(\phi + M)$ or the frequency of models as a fraction of the total number of models and mimics. This expression is essentially the same as one proposed as long ago as 1883 by Blakiston and Alexander (1883). It implies that the predators' reaction to a mimic depends only on their experience of a random sample of the models and mimics which has previously been tested. This is, of course, an extreme over-simplification which may be made more realistic by writing

$$E = \alpha\theta ek + (1 - \alpha\theta)fp \qquad (8)$$

where α is a factor representing the likelihood of mistaken identity of a mimic by a predator, k defines the manner in which subsequent response to a mimic by a predator is modified by previous experience of mimics, and p is the factor by which predation on the mimic compared with the control is increased by its conspicuousness. Both k and α may be expected to be functions of θ and of the intensity of predation. A change in attack on the mimics due to a change in the ratio of mimics to models is described by a change in θ, and a change due to the predators learning that mimics are edible is described by a change in k. The size of α may also vary from one predator species to another depending on the signature by which they recognize the mimics and models. It may increase with time if improvement in mimetic resemblance is effected by selection of genetic modifiers.

Now the instantaneous selection on the two morphs in the edible species may be written as

$$y = 1 - (f/E) \qquad (9)$$

This coefficient represents the advantage to mimics, where the coefficient for the non-mimics is zero. It measures the probability that a predator will attack a mimic compared with a non-mimic when a choice is available. Proceeding as before we next find

$$t = Iy/[(1 - y)(1 - I) + My)]$$

The value of t can then be used in the appropriate difference equation.

If we consider two morphs without specifying their genetic basis this is

$$\Delta M = -tLM/(1 - tM)$$

Before values of ΔM can be calculated it is necessary to consider the type of relation that may exist between frequency and the variables k and α. The factors determining the values of α, the likelihood that the predator will confuse mimic and model, are complex. Duncan and Sheppard (1965) point out that an important factor is the degree of unpleasantness of the model to the predator. When the model is very distasteful or dangerous, confusion of the two is likely; but when mistakes are less disadvantageous, greater discrimination is to be expected. In their example, '. . . Man will be much more cautious in touching the non-poisonous mimic of the deadly coral snake than the harmless mimic of a wasp.' Holling, who discusses the part played by repeated learning and forgetting, has also produced experimental data to show how this modifies the behaviour of vertebrate predators. For the purpose of the present example we shall evade the issue and regard α as a constant. On the other hand, k must certainly vary with mimetic frequency. If every mimic is treated exactly like a model by a predator then k is equal to 1. This behaviour would only occur when $\alpha\theta$ had a large value. If all mimics were treated as edible the appropriate likelihood of predation would be fp, so that $k = fp/e$. This extreme might be met when $\alpha\theta$ was very small. Consequently, the simplest form of regression of k on frequency of mimics is a linear one, for which we obtain the equation

$$k = \frac{fp}{e}(1 - \theta) + \theta$$

Now for a stable equilibrium it is necessary that $t = 0$ at a value of M between 0 and 1. Consequently, from equation (8), E must be equal to f at the equilibrium point, and for stability dE/dM must be positive. Since $e < f$, E can only exceed f if $k > 1$ or $p > 1$. Assume that k and α are both unity, so that at equilibrium

$$E = \theta e + (1 - \theta)fp = f$$

Then

$$\hat{\theta} = \frac{f(1 - p)}{e - fp}$$

so that

$$\hat{M} = \frac{\phi(1 - \theta)}{\theta} = \frac{\phi(e - f)}{f(1 - p)} \qquad (10)$$

With the simplifying assumptions which have been made ϕ, the ratio of models to edibles, is a constant. For example, suppose that models and edibles are in the ratio of 1 : 3, and that non-mimetic edibles are attacked with ten times the frequency of models. We then have $M = -0\cdot3/(1 - p)$, and p must be greater than $1\cdot3$ for equilibrium. If mimics are less than $1\cdot3$ times as conspicuous as non-mimetic edibles the population becomes composed entirely of mimics. Equilibrium rests on the balance between the conflicting requirements of crypsis and resemblance to a conspicuous, warningly coloured, model.

Other ways of maintaining the balance are possible. For example, there may be an antagonism between the protective advantage of a mimetic pattern and an attendant disadvantage during courtship. In butterflies this is an important factor in mimetic polymorphisms, and may account for the prevalence of polymorphisms limited to the female sex (Brower, 1963; Ford, 1964; Magnus, 1963).

Polymorphism may also arise from variation in α or k with the frequency of mimics relative to models, but it is less easy to see how this can transform an advantage of mimicry at low densities into a drawback when density or frequency is high. A more effective agency, if the balance does not depend on the greater conspicuousness of mimics, or on courtship, would be interaction with non-visual pleiotropic effects of the genes involved, which may exhibit heterozygotic advantage, or a non-visual disadvantage of mimics compared to non-mimics (Ford, 1964).

RECIPROCAL SELECTION BY PATHOGEN AND HOST AND A GENERAL METHOD FOR FINDING THE STABILITY OF AN EQUILIBRIUM

When one organism obtains some necessary resource from another, adaptations advantageous to one of the pair may be deleterious to the other. It is possible then for the interaction between the two to give rise to a stable polymorphism. This system may be of importance with respect to the effect of disease organisms on plant and animal populations. New infective strains of organisms such as viruses continually arise to spread through the host populations, causing epidemic diseases. The host population may be severely reduced in numbers, selection for resistant host genotypes takes place along with development of immunogenic and other kinds of resistance on the part of lightly infected individuals, and the incidence of the disease falls. A new outbreak depends on the development of a new infective strain, whereupon the cycle begins again. In man, many detailed case studies are available of epidemics following accidental introduction of a pathogen to susceptible populations where it was previously unknown. The introduction of measles to South America and Polynesia, and of syphilis and bubonic plague to Europe were spectacular examples for which there is little systematic documentation. A great many factors, most of them no doubt social, played a part in the course of these epidemics, but the severity of the disease when spread to a new area indicates that one factor involved is genetic susceptibility. On a less catastrophic scale the succession of influenza epidemics which repeatedly spread through the world illustrates the continuous development of new virulent strains after earlier ones have come under control. A useful example of the natural development of an epidemic in animals is the spread of myxomatosis in rabbits, because no steps were taken to control the disease and, at least in Britain, few attempts were made to spread it. An account is given by Fenner (1965). After decimating the rabbit population, in Australia and in Britain, the pathogen now exists at an equilibrium with a relatively low infectivity among hosts which have increased resistance and modified behaviour patterns. Genetic modification of host or pathogen could now tip the balance one way or the other.

Haldane (1949) was one of the first to discuss the reciprocal selection of pathogens and their hosts in man, but the mathematics of the systems has been developed with particular reference to plants and their diseases. Before considering the conclusions from these studies we shall set up a highly simplified hypothetical system in order to throw some light on the conditions necessary for a polymorphism to exist, but particularly to describe the general method of determining the stability of equilibria in more complex systems.

Suppose we have a system in which a host reproducing asexually is attacked by an asexual pathogen, for example a species of bacterium infected by a bacteriophage. The ecological conditions are such that one species is unlikely to be eliminated by the other. A mutant arises in the host species which is more resistant than usual to attack by the parasite. Let the frequency of this mutant be m, where $l(=1-m)$ is the frequency of typicals. The incidence of exposure to the pathogen is constant, so that the fitnesses of the types with frequencies l and m may be stated as the constants w_1 and w_2. So long as $w_1 < w_2$ the frequency of the mutant will increase.

Now a mutant arises in the parasitic species which is effective in the mutant host but less so in typicals. Let the frequencies of mutant and typical parasites be v and u. If typical and mutant hosts have relative fitnesses w_1 and w_2 when attacked by typical pathogens we may assume for the sake of simplicity that the fitnesses are changed to w_2 and w_1 when infected by mutant pathogens.

To make the model symmetrical the same reversal of fitnesses will apply to the pathogens with respect to the two kinds of host. Then, since a disadvantage to a host type reflects an advantage to a pathogen type we may assign the fitnesses $(2 - w_1)$ and $(2 - w_2)$ to the parasite types at frequencies u and v when infecting typical hosts, and $(2 - w_2)$ and $(2 - w_1)$ when attacking mutant hosts. The changes in frequency in the two species may now be expressed as

$$l_1 = \frac{(w_1 u_0 + w_2 v_0)l_0}{(w_1 u_0 + w_2 v_0)l_0 + (w_2 u_0 + w_1 v_0)m_0}$$

for the hosts, and

$$u_1 = \frac{[(2 - w_1)l_0 + (2 - w_2)m_0]u_0}{[(2 - w_1)l_0 + (2 - w_2)m_0]u_0 + [(2 - w_2)l_0 + (2 - w_1)m_0]v_0}$$

(11)

for the pathogen.

When both polymorphisms are in a state of equilibrium these equations provide

$$\bar{w}_{\text{host}} \Delta l = -lm(1 - 2u)(w_1 - w_2) = 0$$
$$\bar{w}_{\text{pathogen}} \Delta u = uv(1 - 2l)(w_1 - w_2) = 0$$

(12)

A non-trivial equilibrium therefore exists at $\bar{l} = \bar{u} = \frac{1}{2}$. It is now necessary to decide whether it is stable. This is more difficult than in former examples because a change in l is affected by the value of u, and modifies the latter in its turn. The problem is the problem of stability discussed in Chapter 6 with respect to population growth and in this chapter with respect to polymorphism. Now, however, it is extended to two dimensions, l and u. The equilibrium is stable if disturbance from the point $\bar{l}, \bar{u}$ is followed by a tendency to return. If the next and all subsequent movements after the disturbance are away from $\bar{l}, \bar{u}$ then it is certainly unstable, but the path of return is not necessarily direct: it could be curved or even spiral, or it may involve overshoot and return without spiralling in a manner analogous to the damped oscillations described for the population growth equation in Chapter 6 (compare Fig. 4.5). Furthermore, the equilibrium might be stable when frequencies are shifted from it by only a small amount but unstable when larger disturbance is introduced. There are many situations where this difficulty in evaluating equilibria arises. They are ones involving several classes (phenotypes of one or two species), the frequencies of which interact but are partially independent of one another. Examples are some kinds of sex-linked polymorphism, non-random mating systems and all autosomal systems where genotype frequencies are not wholly determined by segregation. In contrast, the stability of the polymorphisms so far discussed is easy to investigate because although there are three genotype frequencies, d, $2h$ and r, they are wholly determined by the gene frequency q through the agency of random-mating and random or genotype-frequency-independent segregation. Were this not so, we would be unable to discuss a polymorphism in terms of Δq alone. At least two equations would be required, e.g. Δd and Δr, because a system of three genotypes would have two degrees of freedom. A general method for investigating stability has been used by several authors (for example, Haldane and Jayakar, 1964; Hull, 1964; Kimura, 1956; Mode, 1958) which involves equa-

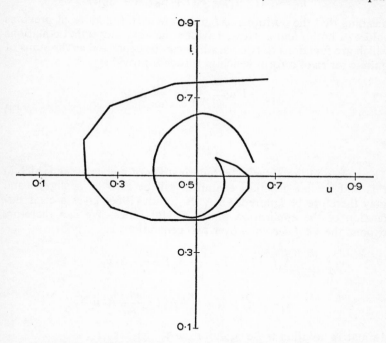

Fig. 4.5. Frequency of a resistance morph in a host species (l) and frequency of a virulence morph in a pathogen (u) in a simple selection system. An equilibrium exists at $l = u = 0·5$. The behaviour over several generations of simulated populations started at $l = 0·55$, $u = 0·55$ shows the equilibrium to be unstable. For the inner curve $w_1 = 0·5$, $w_2 = 1$. The path over 27 generations is shown. For the outer curve $w_1 = 1$, $w_2 = 2$. Twelve generations are shown.

tions for change in frequency of all the components of the system that are not completely determined. It is applicable when relatively small changes are involved. This will be described and applied to the host-pathogen example. The proof of its validity was very kindly provided by Mr Bryan Manly of the University of Salford.

Suppose we start with the system in equilibrium at the point $\bar{l}, \bar{u}$. A disturbance of finite size moves the frequency to the point $l_1, u_1 = (\bar{l} + \Delta_0 l), (\bar{u} + \Delta_0 u)$. The notation Δ_0 represents a change in frequency from the equilibrium point. The type of equilibrium then depends on where $l_2, u_2; l_3, u_3$, etc., lie in relation to $\bar{l}, \bar{u}$. The position in each generation is determined by the frequencies in the preceding one, so that we may write equations of the general form

$$
\begin{aligned}
l_2 &= f_1(l_1, u_1) = f_1(\bar{l} + \Delta_0 l, \bar{u} + \Delta_0 u) \\
u_2 &= f_2(l_1, u_1) = f_2(\bar{l} + \Delta_0 l, \bar{u} + \Delta_0 u)
\end{aligned}
\tag{13}
$$

meaning that the positions of l and u are each functions of previous values of both l and u. Now, Taylor's theorem shows that equations which are functions of two variables can be expressed as the sums of infinite series of a form which in this case provides

$$
l_2 = \bar{l} + \frac{\partial l_1}{\partial l} \Delta_0 l + \frac{\partial l_1}{\partial u} \Delta_0 u
$$

$$
u_2 = \bar{u} + \frac{\partial u_1}{\partial l} \Delta_0 l + \frac{\partial u_1}{\partial u} \Delta_0 u
\tag{14}
$$

There are additional terms with derivatives of higher order, which in this and similar genetical examples will be zero or negligible and may therefore be ignored. C. A. B. Smith (1966) gives a clear discussion of the application of the Taylor series. We can therefore express the *total deviation* over two generations as

$$
\Delta_1 l = l_2 - \bar{l} = \frac{\partial l_1}{\partial l} \Delta_0 l + \frac{\partial l_1}{\partial u} \Delta_0 u
$$

$$
\Delta_1 u = u_2 - \bar{u} = \frac{\partial u_1}{\partial l} \Delta_0 l + \frac{\partial u_1}{\partial u} \Delta_0 u
$$

In matrix notation these equations are

$$
\begin{bmatrix} \Delta_1 l \\ \Delta_1 u \end{bmatrix} =
\begin{bmatrix} \dfrac{\partial l_1}{\partial l} & \dfrac{\partial l_1}{\partial u} \\[2ex] \dfrac{\partial u_1}{\partial l} & \dfrac{\partial u_1}{\partial u} \end{bmatrix}
\begin{bmatrix} \Delta_0 l \\ \Delta_0 u \end{bmatrix}
$$

This may be written in a simpler notation meaning exactly the same thing, as

$$\Delta_1 = M \Delta_0 \qquad (15)$$

where M represents the matrix of derivatives. After a third generation the total change is Δ_2, representing the distance $l_3 - l$, $u_3 - \bar{u}$. This may be found from equation (15) because

$$\Delta_2 = M \Delta_1$$

$$= M^2 \Delta_0$$

Consequently, over $n + 1$ generations

$$\Delta n = M^n \Delta_0 \qquad (16)$$

The requirement for stability of equilibrium is that the right-hand side tends to zero as n tends to infinity. Now, a square matrix raised to the power n can be expressed in the linear form

$$M^n = \alpha_1 \lambda_1^n x_1 + \alpha_2 \lambda_2^n x_2 + \cdots$$

where the α's are constants, the x's are the latent vectors and the λ's are the latent roots of the matrix (see, for example, Smith, 1969, p. 138 *et seq.*). There are as many terms as there are rows and columns in the matrix. M^n can therefore only tend to zero with increase in n if all the latent roots lie within the range $+1$ to -1.

For matrices with two or three rows and columns it is a simple matter to find the latent roots. If we have the matrix

$$\begin{bmatrix} a_{11} & a_{12} \\ a_{21} & a_{22} \end{bmatrix}$$

they are given by

$$\lambda_i = \tfrac{1}{2}[a_{11} + a_{22} \pm \sqrt{((a_{11} - a_{22})^2 + 4a_{12}a_{21})}]$$

For a 3×3 matrix the three roots are found from the equation

$$\lambda^3 - A\lambda^2 - B\lambda - C = 0$$

where

$$A = a_{11} + a_{22} + a_{33}$$

$$B = a_{12}a_{21} + a_{13}a_{31} + a_{23}a_{32}$$

$$- (a_{11}a_{22} + a_{11}a_{33} + a_{22}a_{33})$$

D

$$C = a_{11}(a_{23}a_{32} - a_{22}a_{23})$$
$$- a_{12}(a_{21}a_{33} - a_{31}a_{23})$$
$$- a_{13}(a_{31}a_{22} - a_{21}a_{32})$$

When the matrix is larger it is more tedious and less useful to find latent roots in algebraic terms, but they may be evaluated numerically for specific cases. It may be noted that only the largest latent root is required and that this one is found first by the method of evaluation used on computers. We therefore have a general method for investigating stability, the first steps of which are to find recurrence equations describing the change in frequency for each partially independent component and to differentiate them with respect to each component in turn.

In the host-pathogen model we have

$$l_1 = (w_1 u_0 + w_2 v_0)l_0 / \overline{w}_{\text{host}}$$
$$u_1 = [(2 - w_1)l_0 + (2 - w_2)m_0] u_0 / \overline{w}_{\text{pathogen}}$$

Differentiating and substituting the equilibrium values $l = \bar{u} = \frac{1}{2}$ provides the matrix

$$\begin{bmatrix} 1 & \dfrac{w_1 - w_2}{w_1 + w_2} \\ \dfrac{w_2 - w_1}{4 - w_1 - w_2} & 1 \end{bmatrix}$$

for which the latent roots are

$$\lambda_i = 1 \pm \sqrt{\frac{-(w_1 - w_2)^2}{(w_1 + w_2)(4 - w_1 - w_2)}}$$

There can therefore be no values of w_1 and w_2 giving $|\lambda_1|, |\lambda_2| < 1$, so that the system is unstable. The nearest approach occurs when $w_1 = w_2$, whereupon both roots equal 1, the equilibrium is neutral and the frequencies, if disturbed from equilibrium, remain at the point to which they are moved. Two paths of successive values of l and u are illustrated in Fig. 4.5.

The stability of simple single locus polymorphisms has been discussed in terms of difference equations. It would be consistent to do the same for the general treatment. For this purpose $l + \Delta_0 l$ is substituted for l_1 and $u + \Delta_0 u$ for u_1. The matrix of partial derivatives then becomes

$$\begin{bmatrix} \dfrac{\partial(l + \Delta l)}{\partial l} & \dfrac{\partial(l + \Delta l)}{\partial u} \\[3ex] \dfrac{\partial(u + \Delta u)}{\partial l} & \dfrac{\partial(u + \Delta u)}{\partial u} \end{bmatrix} = \begin{bmatrix} 1 + \dfrac{\partial \Delta l}{\partial l} & \dfrac{\partial \Delta l}{\partial u} \\[3ex] \dfrac{\partial \Delta u}{\partial l} & 1 + \dfrac{\partial \Delta u}{\partial u} \end{bmatrix}$$

The subscript 0 has been dropped from the Δ's to simplify the notation: equilibrium values of l and u should be inserted in each difference equation. As before, the absolute values of the latent roots of the matrix must be less than 1. It may then be shown that if we use the simple matrix

$$\begin{bmatrix} \dfrac{\partial \Delta l}{\partial l} & \dfrac{\partial \Delta l}{\partial u} \\[3ex] \dfrac{\partial \Delta u}{\partial l} & \dfrac{\partial \Delta u}{\partial u} \end{bmatrix}$$

the latent roots λ'_i must lie between 0 and -2, so that $(\lambda'_i + 1)^n$ tends to zero as n increases. For a 2×2 matrix they will be negative if (Hull, 1964) $a_{11} + a_{22} < 0$ and $a_{11}a_{22} - a_{12}a_{21} > 0$. For a 3×3 matrix Mode (1958) shows the equivalent condition to be that $A < 0, B < 0, C < 0$ and $AB - C > 0$, these values being as defined above. The equations for the host-pathogen example provide

$$\begin{bmatrix} 0 & \dfrac{w_1 - w_2}{w_1 + w_2} \\[3ex] \dfrac{w_2 - w_1}{4 - w_1 - w_2} & 0 \end{bmatrix}$$

illustrating clearly the relationship between the two methods. The roots are

$$\lambda'_i = \pm \sqrt{\frac{-(w_1 - w_2)^2}{(w_1 + w_2)(4 - w_1 - w_2)}}$$

There is therefore no real part between zero and -2.

The haploid host/haploid pathogen model leads to an unstable equilibrium. It is unlikely that polymorphism could arise between a pair of haploid organisms unless fitnesses show more complex frequency dependence than is built into these equations. Interest in the possibility of polymorphisms of a related kind has arisen from the work of H. H. Flor on plant hosts such as wheat and their obligate fungal parasites, which cause rust and smut diseases. Flor and other workers have shown that there are related polymorphic loci in the hosts and their parasites, the host locus determining resistance or susceptibility to attack and the locus in the parasite governing its virulence or avirulence. Many loci may be involved. The resistance genes of the host are multiple alleles, while the virulence genes in the disease organisms, specific to particular alleles of the host, are at separate and often unlinked loci. The relationship of the genes is operational but not necessarily structural. Person, Samborski and Rohringer (1962) define the system as one where 'the presence of a gene in one population is contingent on the continued presence of a gene in another population, and where the interaction between the two genes leads to a single phenotypic expression by which the presence or absence of the relevant gene in either organism may be recognized'.

Mode (1958) developed a mathematical theory which allows stable equilibria. It recognizes two important biological properties of the interacting species. First, the fungal parasites, like their hosts, may be treated theoretically as diploids, since infection occurs during the diploid para-sexual stage. Secondly, the resistance genes of the host are dominant to non-resistant alleles—when two resistant alleles are present together they are both active—while genes for virulence in the pathogens are recessive to their avirulent alleles. In the algebraic model there are two resistance alleles and a non-resistant allele in the host, which is assumed to be an outbreeder, and virulence genes specific to each of them at unlinked loci in the parasite.

Proceeding in a similar manner to that used here, Mode derives conditions for stable equilibrium. The important feature of the system with respect to polymorphism is that heterozygotes for both resistance genes are resistant to a wider range of pathogens than either homozygote, so that heterosis is involved. This point is emphasized by Person (1966), who also implies that the dependence of the fitness of each species on the frequency of genotypes in the other species will favour polymorphism, likening the system to Batesian mimicry. It is difficult to see that this can be so without the addition of heterosis, so that establishment of polymorphism requires the correct dominance relations in host and parasite. Once these are available for one host and parasite locus, however, Person shows that the way is open for increasing complexity of interaction, the addition of another resistance allele at the host locus selecting for virulence alleles at another parasite locus. Although the host species are now inbreeders it is probable that the multi-allele/multilocus systems that have been studied arose in this way.

5

POLYMORPHISM IN NON-PANMICTIC
CONDITIONS

MIGRATION

The true panmictic unit, in which there is random mating and an equal opportunity for any combination of mates to arise, must be rare. When populations are large there are ecological and geographical restrictions on the available choice of mates. When they are small there may be restrictions associated with factors such as family structure. Species are often composed of a series of small semi-isolated units, in each of which there is a degree of inbreeding. The effect of partial isolation on the variance in gene frequency over an area has been examined in a series of papers by Sewall Wright, and is dealt with by Li (1955a) and Moran (1962). If selection differs from place to place the result of diffusion from one part of the range to another may be continued genetic heterogeneity over wide areas.

The rate of migration

The migration rate from one place to another may be expressed in several ways. It may be measured as the fraction m of the breeding adults of a colony in each generation which is replaced by immigrants from other colonies. This is a greatly oversimplified way of looking at animal movement. In nature there is often continuous diffusion, some animals covering long distances from their birth places, while others hardly move at all, or perhaps travel on circular paths to return eventually to their places of origin. The pattern of distribution may not consist of colonies with distinct boundaries across which the migrants move. In reality the nearest approach to a boundary between breeding units may be an area of low density where it is difficult to assign individuals to one group or the other. Nevertheless,

when attempting to obtain a picture of selection in natural populations the colony and migrant model is probably almost always a good
enough description of the relatively inexact data that are available.

With respect to the biological meaning of a value m, it may be
noted that if we consider two populations of equal size where each
individual of one has an equal probability of mating with a member
of its own or of the neighbouring group, then $m = 0.25$. This is
therefore the value equivalent to random mating within a single
population. When two populations of approximately equal size are
recognizably separate and distinct m will lie somewhere between
zero and 0.25. If numbers are very unequal, however, this limit does
not apply and the breeding population in a small group surrounded
by large ones could include up to 50% of immigrants on the assumption of migration rates at no more than the random mating level.

Consider two sub-populations of sizes N_i and N_j. Their relative
sizes are the fractions $c_i = N_i/(N_i + N_j)$ and $c_j = N_j/(N_i + N_j)$. Each
is composed of 50% males, and for convenience we may say that the
males migrate while the females remain stationary. For random
mating a male of population i has a chance c_i of mating with a
female of its own population, and c_j of mating with a female from
the other group. In population j the chances are c_j and c_i. The frequency of matings between indigenous females and migrants is
accordingly $c_i c_j$ in each group, and the migration rate from i to j is
$\frac{1}{2}c_j$; i.e. that fraction of N_i leaves to be replaced by $\frac{1}{2}c_j$ immigrants.
With respect to the N_j in population j, a fraction $\frac{1}{2}c_i$ leaves and is
replaced by immigrants. If $N_i = 7000$ and $N_j = 3000$, then under
conditions of random mating 15% of N_i and 35% of N_j are immigrants in each generation, these fractions being 1050 individuals. An
observed rate of migration can therefore be compared with a standard
representing a certain kind of behaviour.

Another way of assessing the redistribution of individuals when
there is movement between several colonies of unequal size has been
used by Deakin (1966). He defines k as the fraction of gametes from
each colony which leave it to be redistributed at random, so that a
proportion arrive back in their original location. Suppose that there
are 200 individuals in niche i and 1800 in niche j. With equality of
the sexes and random mating, the 100 males in i have a 10% probability of mating with indigenous, and a 90% probability of mating
with exogenous females. The 90 males which leave are replaced by
90 immigrants. The migration rate in colony i is therefore $90/200 =
m = \frac{1}{2}c_j$. In colony j it is similarly $10/200 = m' = \frac{1}{2}c_i$, the m's being
the fractions of the colonies referred to which consist of immigrants.
In terms of k we may say that 50% of gametes result from matings
inter se within a niche, and 50% from matings at random between

individuals from both niches, so that $k = \frac{1}{2}$. In niche i 100 adults stay put, while 100 mate at random with the 900 vagrants from the other population. One tenth, or c_i, of the group return to i, including 10 individuals originally from i and 90 from j. The fraction leaving to undergo random union is k, and the 110 indigenous individuals in niche i after the reorganization has taken place comprise $(1 - k)$ $c_i + kc_i^2$ of the total. This fraction is $(1 - m)c_i$, so that the three factors are related by the equation $k = m/(1 - c_i)$.

Reference to a migration rate equivalent to random mating does not mean, however, that the pair of populations with migration rates $\frac{1}{2}c_i$ and $\frac{1}{2}c_j$ will have genotype distributions identical with that of a single random mating population. There is a difference unless the mean gene frequency is the same in all samples taken from different parts of the interbreeding area. If q_i and q_j are not identical at the start, the difference can only be overcome in one generation by migration rates of twice the random mating level. With lower rates of migration in the absence of selection, the difference between the frequencies in the two populations will be reduced in successive generations at a rate depending on the value of m. After one generation the gene frequencies are

$$q_{i1} = (1 - m)q_{i0} + m\, q_{j0}$$

$$q_{j1} = (1 - m)q_{j0} + m\, q_{i0}$$

so that $(q_{i1} - q_{j1})/(q_{i0} - q_{j0}) = 1 - 2m$ is the factor by which the difference decreases per generation.

Suppose there is no migration between two colonies of relative sizes c_i and c_j. The overall mean frequency is $\bar{q} = c_i q_i + c_j q_j$; and if the populations are reasonably large the genotype frequencies in each are $p_i^2 : 2p_i q_i : q_i^2$ and $p_j^2 : 2p_j q_j : q_j^2$. Now if subdivision had no effect the genotype frequencies estimated from $\bar{q}$ should be identical with the mean of the genotype frequencies for each colony. But considering only heterozygotes for the moment, half their mean frequency is

$$h_m = c_i p_i q_i + c_j p_j q_j$$

where the estimate of h from $\bar{q}$ is

$$h_t = (c_i q_i + c_j q_j)(1 - c_i q_i - c_j q_j)$$

$$= c_i p_i q_i + c_j p_j q_j + c_i c_j (q_i - q_j)^2$$

Therefore, $h_t \neq h_m$, and since the third term in the last equation must be positive, $h_t > h_m$. The two values of h can only be equal if $q_i = q_j = \bar{q}$, which would require migration rate of c_i and c_j, or twice the 'random mating' level. The amount by which heterozygotes are reduced is in fact the variance of the difference in gene frequency between populations, this amount being added on equally to the two homozygous classes. For a systematic account of the effects of subdivision the reader is referred to ch. 21 of Li (1955a).

Two populations of unequal size

The populations of which a species is composed are often very different in size. The inequality may reflect variability in the level of selective pressures, small populations living in extreme environments and large ones in more benign ones. The effect of gene flow from the small to the large populations is negligible, while the reverse flow may have a stultifying effect on the response shown to selection by the small marginal unit (e.g. see Mayr, 1963; Birch, 1960).

The relation is expressed simply as follows. In the small population three genotypes, AA, Aa, aa, have frequencies, p^2, $2pq$, q^2, with fitnesses, $(1-s)$, 1, $(1-t)$. In the large population the frequency of gene a remains constant at q_m, and in each generation there is migration from the large to the small colony at rate m. The value m is the fraction of the breeding adults that are immigrants.

The precise change in gene frequency due to a combination of selection and immigration depends on the point in the life cycle at which gene flow occurs. If immigrants arrive before the selection which determines the final adult frequency, then of course they are involved in it and suffer according to their genetic constitution. If they arrive after selection they simply contribute a fraction of the gametes which form the next generation. The difference this makes is akin to the difference between simultaneous and successive selection.

Assuming arrival after selection we may write the recurrence equation

$$q_1 = \frac{(1-m)q(1-tq)}{1-sp^2-tq^2} + mq_m \tag{1}$$

where the first term is the change under selection of the indigenous members, and the second is the contribution of immigrants. From it we get

$$\Delta q = \frac{pq(sp-tq)-mq(1-tq)}{\overline{w}} + mq_m \tag{2}$$

which describes the general relation between selective values, migration and frequency.

If there is no selection the change in frequency resulting from migration may be described simply by the equations,

$$q_1 = (1 - m)q_0 + \quad q_m$$

and

$$\Delta q_{(m)} = m(q_m - q) \tag{3}$$

Note that like mutation, but unlike selection, the change in gene frequency in a population is not of necessity zero when q equals 0 or 1. Setting $m = 0$, equation (1) gives the change in frequency in the absence of migration. Call this value $\dot{q}_1$. Now when migration occurs equation (1) is

$$q_1 = \dot{q}_1 - m\dot{q}_1 + mq_m$$

and (2) is

$$\Delta q = \dot{q}_1 - q_0 + m(q_m - \dot{q}_1)$$
$$= \Delta q_{(s)} + m(q_m - q_0 - \Delta q_{(s)})$$
$$= (1 - m)\Delta q_{(s)} + \Delta q_{(m)} \tag{4}$$

where $\Delta q_{(s)}$ is the change due to selection alone. When m is small the overall change is almost the sum of the two independent terms. Returning to the problem of migration from a large to a small population we may consider the extreme case where $t = 0$ and q is the frequency of a new mutation which has arisen in a small population. The mutation confers a slight advantage, estimated from the relative disadvantage s, and it is absent from the large neighbouring population. Equation (2) reduces to

$$\Delta q = sp^2 q - mq$$

so that $sp^2 - m = 0$. The disadvantage of the typical compared to the new dominant mutant must therefore be greater than the fraction of immigrants to allow advance in frequency.

Wright (1940) and Li (1955a) discuss the situation where q_m is the mean equilibrium frequency in the neighbouring populations providing immigrants. If it is assumed that there is no dominance and that the value m is small, so that $(1 - m)$ is near unity, we have

$$\Delta q = \Delta q_{(s)} + \Delta q_{(m)}$$
$$= tq^2 - q(t + m) + mq_m = 0$$

from which

$$\hat{q} = \frac{t + m}{2t} \pm \sqrt{\left(\frac{(t + m)^2}{4t^2} - \frac{mq_m}{t} \right)} \tag{5}$$

As an example, Fig. 5.1 shows the curves for change under selection and migration when $s = 0.2$, $t = -0.2$, $m = 0.1$ and q_m is constant at 0.3. The overall Δq curve is the sum of the other two. Using equation (5) to give an indication of the equilibrium point, even though s and m are large, we find $\bar{q} = 0.7$ or $\bar{q} = -0.2$. In the figure it can be seen that the resultant curve cuts the q axis in both these regions.

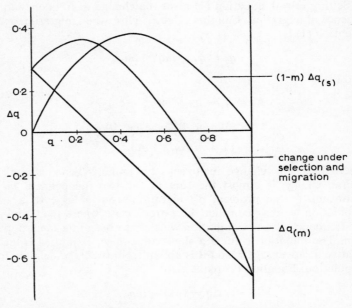

Fig. 5.1. Combination of selection and migration. $s = 0.2$, $t = -0.2$, $m = 0.1$, $q_m = 0.3$. For explanation see p. 105.

Two populations of similar size

When populations are more or less similar in size and partially isolated from each other there is a possibility of reciprocal migration between them. If selection is in different directions in the two environments its action is said to be disruptive (Mather, 1953). With reduction in the migration rate or increase in inviability of the F_1 of migrant/indigenous matings, selection under these conditions may ultimately lead to speciation, but if gene flow prevents complete separation the result is polymorphism.

Working as in equations (2) to (4), and using subscripts i and j to represent frequencies and coefficients in the two populations, we have

$$q_{i1} = (1 - m)\mathring{q}_{i1} + m\mathring{q}_{j1}$$
$$q_{j1} = (1 - m)\mathring{q}_{j1} + m\mathring{q}_{i1}$$
(6)

with migration at the same rate in the two directions, and

$$\Delta q_i = (1 - m)\Delta q_{i(s)} + m\Delta q_{j(s)} + \Delta q_{(m)}$$
$$\Delta q_j = (1 - m)\Delta q_{j(s)} + m\Delta q_{i(s)} - \Delta q_{(m)}$$
(7)

In both equations the change due to migration is $\Delta q_m = m(q_j - q_i)$
When $\Delta q_i = \Delta q_j = 0$, equations (7) can be written

$$-\Delta q_{i(s)} - m\Delta q_{j(s)} + m\Delta q_{i(s)} - \Delta q_{(m)} = 0$$
$$\Delta q_{j(s)} - m\Delta q_{j(s)} + m\Delta q_{i(s)} - \Delta q_{(m)} = 0$$

so that at frequencies $\mathring{q}_i, \mathring{q}_j$

$$\Delta q_{j(s)} = -\Delta q_{i(s)}$$
(8)

This is a necessary requirement for equilibrium.

Since there are involved four selective coefficients, two equilibrium frequencies and a migration rate, the equilibrium conditions cannot be stated concisely. The picture is further complicated if the migration in the two directions is unequal. To give the simplest example, suppose there is no dominance, so that $s_i = -t_i$ and $s_j = -t_j$, and further, that selection is acting with equal force but opposite direction in the two colonies (i.e. $s_i = t_j$). Letting $\bar{w}_i = \bar{w}_j \simeq 1$, equation (8) provides

$$p_i q_i = p_j q_j$$

Consequently, at equilibrium $p_i = p_j$ or $p_i = q_j$. Substitution of $p_i = p_j$ in the first of the two equations (7) provides

$$(2m - 1)tp_i q_i = 0$$

which can only be true over a range of values of migration if $p = 0$ or 1. Substitution of $p_i = q_j$ in the same equation gives

$$q_i^2 s(2m - 1) + q_i(s - 2ms - 2m) + m = 0$$

so that

$$\mathring{q}_i = \frac{1}{2} + \frac{m}{s(2m - 1)} \pm \sqrt{\left[\frac{1}{4} + \left(\frac{m}{s(2m - 1)}\right)^2\right]}$$
(9)

If we take s to be positive and $m < \frac{1}{2}$, the terms in m and s are zero when $m = 0$ and negative for greater values of m. The sign in front of the radical therefore has to be positive, since a negative sign gives values of q of 0 or less.

Differentiating the Δq equation we have

$$\frac{d\Delta q}{dq_i} = 4q_i\, ms - 2q_i s + s - 2ms - 2m$$

Since this is negative for all relevant values of m and s the equilibrium is stable. The conditions are restricting, however, because a precise balance between the selective pressures in the two populations has been assumed, and is necessary to maintain the equilibrium. As an example, suppose that $m = 0.1$ and $s = 0.2$. We then have $\hat{q}_i = 0.5$ $- 0.625 + \sqrt{0.641} = 0.675$. The value of $\hat{q}_j$ is therefore $1 - 0.675 = 0.325$. The change under selection at $\hat{q}_i$ in population i is

$$\Delta q_{i(s)} = \frac{0.2\hat{p}_i\hat{q}_i}{1 - 0.2(1 - 2\hat{q}_i)} = 0.041$$

and at q_j

$$\Delta q_{j(s)} = \frac{-0.2\hat{p}_j\hat{q}_i}{1 + 0.2(1 - 2\hat{q}_j)} = -0.041$$

If s is only 0.02 instead of 0.2 the equilibrium frequencies are $\hat{q}_i = 0.52$ and $\hat{q}_j = 0.48$.

The greater degree of complexity of this situation compared with the previous one in which q_m was a constant is clear when the components of Δq are graphed. The changes under selection may be expressed in the same way as in Fig. 5.1, but we cannot include Δq_m in order to read off an equilibrium, because it is a function of q_i and q_j, which interact with each other. If q_i is plotted on q_j the values of Δq_m form an inclined surface falling from a value of m at $q_j = 1$, $q_i = 0$ to $-m$ at $q_i = 1$, $q_j = 0$. The diagonal $q_i = q_j$ is a contour with the value $\Delta q_m = 0$. In the absence of selection a population starting at any other point on the surface on either side of the line will approach it along a straight path at right angles to it, the speed of travel depending on the migration rate. Fig. 5.2(a) shows this graph with the path from $q_j = 0.9$, $q_i = 0.1$. In 5.2(b) the course of events in time is shown. The dotted lines represent the approach to equilibrium when there is no selection and $m = 0.1$. The other two curves are the paths followed by the two populations under the conditions of the numerical example, starting from the frequencies 0.1 and 0.9. The final equilibrium state depends on the balance of opposing selection pressures and the flow of disadvantageous genes, which in turn is affected by migration rate and the type and size of selection.

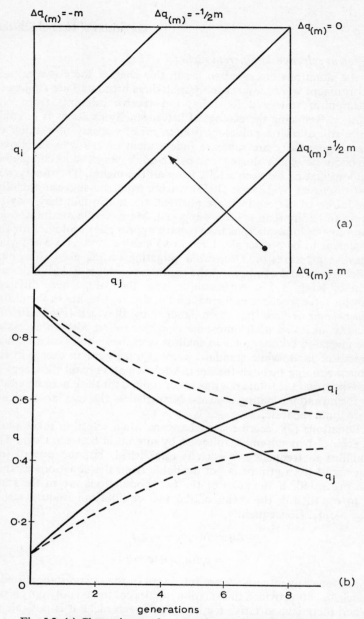

Fig. 5.2. (a) Change in gene frequency in two colonies as a result of migration in the absence of selection. The arrow shows the direction of change from $q_j = 0 \cdot 9$, $q_i = 0 \cdot 1$. The diagonal lines are contours of equal value of $\Delta q_{(m)}$. (b) Change in gene frequency with time in two colonies in the absence of selection (dotted line) and under the combination of migration and selection described on p. 108. The dotted lines describe the same process as the arrow in (a).

Different selection in different niches

In the situations discussed so far in this chapter there is a mosaic environment where several sub-populations breed and are subject to selection in restricted localities, but receive migrants from one another. Reducing the ecological heterogeneity as far as is possible while still allowing polymorphism to arise, we may now think of cases where there are separate independent niches with their own selection pressures during some pre-adult stage, but completely random mating between adults from all the niches. The theory was first discussed by Levene (1953). In his paper he suggests that the conditions of life outlined are not realistic, but in fact they may be closer to reality than is often supposed. Many sessile marine species and most holometabolous insects have appropriate ecologies for the argument to be applicable. Even in so small a unit as a 3×1 vial, selective pressures on *Drosophila* pupating on the glass sides may differ from those of pupae in the medium (compare Sokal in discussion of King, 1955). An example from the wild, where different chromosome frequencies were found in different niches in the midge *Chironomus tentans* has been studied by Blaylock (1965). In one locality an inverted chromosome type was found with an excess of the inversion homozygotes in shallow man-made holes on a marshy stretch of land, while standard homozygotes were in excess in the stream running through the area. Adult females lay all their eggs at one time, and the total area was small enough for the emerging adults to form random mating swarms. Nevertheless the two larval niches were quite separate.

Equations (7) describe what happens when selection takes place in each of two sub-units, followed by migration between them. The equilibrium frequency in each is established, but the overall frequency of both groups is not available from the equations as they are presented. If the sizes of the two populations are in the ratio c_i to c_j, then $\bar{q}$, the mean of the two equilibrium frequencies, is $c_i q_i + c_j q_j$. Consequently,

$$\Delta \bar{q} = \Delta(c_i q_i + c_j q_j)$$
$$= c_i \Delta q_{i(s)} + c_j \Delta q_{j(s)} \tag{10}$$

On Levene's hypothesis we are interested in the circumstances under which $\Delta \bar{q} = 0$, provided this relation implies continued polymorphism. When there is assortative mating within each niche it is possible to have a situation where the sum of the weighted changes in frequency $c_i \, \Delta q_{i(s)}$ and $c_j \, \Delta q_{j(s)}$ is zero because gene frequencies are diverging from the mean at an equal rate. This is true, for example, after generation 5 in Fig. 5.2(b). Discussion of change in $\bar{q}$ is therefore applicable if and only if all adults mix together, so that the frequency in each niche is the same before selection and genotypes are initially in the Hardy–Weinberg ratio.

When these conditions hold the argument goes on as follows. Using i to denote one out of an array of n niches including j, equation (10) may be rewritten

$$\Delta \bar{q} = \sum_{i=1}^{n} \frac{c_i p_i q_i (s_i p_i - t_i q_i)}{\bar{w}_i}$$

$$= pq \sum_{i=1}^{n} \frac{c_i (s_i p_i - t_i q_i)}{\bar{w}_i} \tag{11}$$

so that there are equilibria at $q = 0$, $q = 1$ and at any point or points where

$$\sum c_i (s_i p_i - t_i q_i) = 0 \tag{12}$$

Non-trivial equilibria may be shown to be stable by demonstrating that Δq can be positive for very small q and negative for very large q: conditions which enable Δq to cross the $\Delta q = 0$ line with a negative slope at some intermediate point. This is done by substituting $p = 1$ and $q = 1$ successively into (12), providing the requirement that $\sum c_i s_i$ and $\sum c_i t_i$ both be positive. There is then overall heterozygote advantage, so that no novel route to a stable equilibrium has so far been established by the introduction of the hypothesis of several niches. The conditions are the same as those discussed on pp. 73–5, where the joint action of different selective agents or of selection acting on different pleiotropic effects is considered, except that previously the factor c_i was included in the calculation of Δq. As Levene points out, however, the stated conditions are not the only ones to provide stable equilibria, although they will certainly do so.

Fig. 3.1 (p. 61) shows the path of Δq on q under different selective regimes when there are two niches. In 3.1(a) the conditions are, for the first niche, $c_1 = 0.4$, with selective coefficients 0.2 and -0.5, and for the second, $c_2 = 0.6$ with coefficients -0.1 and 0.4. There is mean heterozygote advantage, and the curve is the one that would be obtained using the coefficients 0.02 and 0.04 (that is, the means of the s_i and t_i) in the elementary equation. On the other hand, Fig. 3.1(b) shows another two-niche situation. Here, Δq is positive at both extremes but a stable equilibrium occurs because the zero line is cut twice, giving rise, in addition, to an unstable equilibrium. (The values are $c_1 = 0.6$, with selective coefficients -1 and -0.1, and $c_2 = 0.4$ with coefficients 0.5 and -0.1. The selective pressures are used by Levene in one of his examples.) There is no longer mean heterozygote advantage, $\sum c_i s_i$ being -0.4 and $\sum c_i t_i$ being zero, and the balance is not attributable to heterosis as such but to a conflict between selection and migration in each niche. In the same way, the equilibria discussed on pp. 105–8 arise from conflict between selection and migration without heterosis necessarily being involved. It seems unlikely that such conditions, where there is heterozygote disadvantage in one niche but directional selection in the other can arise at all commonly in nature, but the theoretical consequences are interesting. Li (1955b) has shown in an elegant way that a general statement of the conditions for equilibrium may be made by defining $\bar{w}$ for the random mating population in n niches as $\prod_{i=1}^{n} \bar{w}_i^{c_i}$. This function maximizes at stable equilibria, just as for the simple panmictic system the mean fitness $\bar{w}$, as usually defined, maximizes at stable equilibria.

The conditions for stable equilibrium are also discussed by Deakin (1966). As mentioned above, the value k is defined as the fraction of gametes which may be thought to leave their niche and be relocated at random, including return to the original niche. In Levene's analysis $k = 1$, while lower values would occur if there was a tendency for some adults emerging in a niche to lay eggs preferentially in it. After some algebra Deakin shows that for small changes sufficient conditions for polymorphism are that $s_i > k(1 - c_i)$ and $t_j > k(1 - c_j)$ for some populations i and j in the series. The values $k(1 - c_i)$ and $k(1 - c_j)$ are equal to the factors m for the two populations, that is, to the fraction of individuals which are immigrants in populations i and j. This conclusion may be compared with the conclusion following equation (4), that the selection coefficient favouring an allele and migration of individuals carrying the other allele approximately balance one another.

The mathematics of systems where there is differential selection between niches, with and without habitat selection by the adults raised in them, is discussed in an extensive paper by Maynard Smith (1966), who goes on to consider the possibility that such selection will result in sympatric speciation. He concludes that it may do so, but only under restricted conditions, the initial polymorphism that depends on niche heterogeneity being relatively easily put into a state of imbalance by small fluctuations in the forces involved, unless selective coefficients are large.

Several populations of similar sizes

The conclusions reached for two populations are valid when extended to a large number of units. Sewall Wright has stressed the importance of this kind of distribution of species, where semi-isolated colonies are subjected to diverse selective pressures, in promoting rapid evolution. Conditions of inbreeding in the small units heighten the divergences between them, raising the variance of gene frequency about the overall mean to a level that would not be attained if there were larger areas of panmixia. Coupled with this variability, the selection pressures themselves vary in space. Even when there is not a tendency to a common stable equilibrium, it is certain that the tendency to homozygosity at any locus is very much slowed down in such arrays. This may be an important factor maintaining the widespread polymorphism in some animal groups such as helicid snails —the most intensively studied of which is *Cepaea nemoralis*—where selective pressures certainly fluctuate greatly in direction and magnitude over short distances. Provided there are some places where selection maintains a stable polymorphism in the species by means which involve small areas, there could be a good deal of subordinate polymorphism in adjacent localities which, if completely isolated, would become monomorphic.

There are two formal approaches to the situation, called respectively the 'island' or 'stepping stone' model and isolation by distance. The one employed falls into the first category. Each unit is treated as if it were an entity with its own frequencies and selection pressures but contributing a certain proportion of its members to the next unit. The sequence may be extended in one or two dimensions. In one dimension we can consider a chain of units with migration taking place to the adjacent unit but not further down the chain, and with some systematic selective change along the length. Equations like (6) and (7) would hold for each colony, but the immigrants would have a frequency which was the average of those of the populations on each side. If each population receives immigrants at rate m from n adjacent colonies, the equation describing its change in frequency is

$$\Delta q = (1 - m)\Delta q_{(s)} + \frac{m}{n} \sum_{a=1}^{n} \Delta q_{a(s)} + \Delta' q_{(m)}$$

The last term represents the mean effect of immigrants, being

$$m\left[\frac{1}{n}\left(\sum_{a=1}^{n} q_a\right) - q\right]$$

This kind of model has two virtues. The simple linear pattern is relevant to a natural situation commonly encountered along ecological edges, the most striking being littoral and sublittoral environments but including such features as hedgerows, walls and cliff exposures. It is also simple enough to allow simulations to be run easily on the computer. As a description of the true conditions under which animals live it is naïve. In reality each individual has a probability of migration over a given distance, the probability distribution often being markedly leptocurtic, and at the small scale the kind of family unit influences the degree of randomness of mating. More elegant mathematical treatments take such factors into account, but in considering them we have to keep the present objective in mind, which is to understand the effect of the principal factors involved, and where possible to arrive at estimations of selection. For these purposes simulation on the basis of the island model should usually be perfectly satisfactory, allowing as it does adjustment of selection conditions and migration rates from place to place. In addition, the model may easily be extended to two dimensions in a computer simulation. Theoretical models of the alternative type are discussed by Wright (1946 and elsewhere), and Moran (1962). Haldane (1948) dealt with a continuous linear model that has been applied to real data by Haldane himself, and by Kettlewell and Berry (1961). Selection is considered to change in direction at a demarcation line, while diffusion of individuals takes place across the line.

In Haldane's example he uses the results of Sumner and Blair, who studied morph frequency in the deer-mouse *Peromyscus*. In the area of study there are light-coloured subspecies on sandy beaches by the coast carrying a dominant lightening gene, which are replaced by the darker typical subspecies on blacker soils inland. It is easy to visualize a reversal of selective pressure at the sharp boundary between the two zones while migration across it is not unduly restricted.

The example studied by Kettlewell and Berry is the cline of frequency of melanics in the moth *Amathes glareosa* in Shetland. Here, there is a marked drop in morph frequency at a natural feature, the Tingwall valley, which may be a barrier to movement. It may also coincide with a discontinuity in selective pressure, but this is not certain. The starting assumptions are therefore not identical to those of Haldane.

Fisher (1950) discusses a case where there is no dominance, while selective pressure changes as a linear function of distance along the length of the territory involved. The combination of progressive change in selection with diffusion of genes gives rise to a sigmoid curve of gene frequency symmetrical about the mid-point. The steepness of the change in this, as in Haldane's, cline depends on the relation between change in selection and rate of diffusion. In both papers details and tables are given for fitting the curves to empirical results. Using either method, or a modification of them to suit a particular situation, estimates of the relation of selective coefficients to average movement can be made. If there is independent evidence on either parameter absolute estimates may be obtained.

In the studies of both Kettlewell and Berry and of Haldane the rate of movement can be assessed to within about a factor of two. The analysis provides a difference between the selective coefficients in the region where the gene is favoured and where it is at a disadvantage of no more than 3%, and probably much less. This is a net estimate and may imply that there is a balance between advantageous and deleterious effects of the genes. In the study of the moth there is certainly evidence for visual selection of a higher order of magnitude.

Despite the fact that the method has been used successfully to provide estimates of coefficients, it seems less satisfactory than simulation on basis of an island model. The assumptions for the curve are bound to be extreme oversimplifications, and in order to obtain tractable equations the parameters must remain constant over the length of the curve, or change according to very simple laws. By contrast, even a simple-minded computer simulation can incorporate local fluctuations in conditions. The fineness of scale of simulation can be altered by raising or decreasing the number of units involved, and a degree of smoothing which may approach true ecological conditions more closely can be introduced by allowing a fraction of migrants to travel in one generation to colonies at any chosen distance beyond the adjacent ones. In Fig. 5.3 two curves for

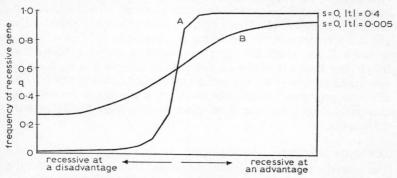

Fig. 5.3. Result of 1000 generations of selection and migration across a boundary in a 20-unit computer simulation. To the left of the mid-point the fitnesses are $1 : 1 : (1 - t)$; to the right they are $1 : 1 : (1 + t)$. For all units except the two terminal ones, 5 per cent of each migrate to each adjacent one per generation. At the ends there is only 5 per cent migration in all. For curve A, $|t| = 0.4$; for curve B, $|t| = 0.005$.

computer outputs based on a 20-island chain are shown. In each, selection is acting on the recessive genotype, and the selective coefficient reverses in sign at the midpoint. The value of m is 0·1. For curve A the selective coefficient t has a value of 0·4 to the left and

-0.4 to the right; for curve B the coefficients are 0.005 and -0.005. Both simulations were run for a thousand generations, starting with a gene frequency of 0.5 throughout the range. The final pattern was in fact reached in about 35 generations in curve A but required several hundred generations to become fully established with the much smaller selective pressures of curve B. It would have been attained whatever the starting frequencies so long as directional selection did not lead to extinction of one of the genes before it was present on both sides of the boundary.

These curves are based on the same assumptions about selection as Haldane's but movement of individuals occurs in discrete jumps on the computer, while it is normally distributed in his analysis. In the computer model 10% of individuals move one unit of distance from their starting point in each generation. In order to compare the result with that of Haldane we therefore have to standardize the distances involved with respect to this fraction of movement. If 10% of individuals find themselves beyond a deviate of 1 unit in a normal distribution then the mean squared deviation of the distribution is 0.61. This is a value called m by Haldane, which we may rename μ to avoid confusion with the migration rate already defined. The abscissa of his figure is graduated in terms of μ, and the relation of selection pressure to migration rate is discussed in terms of a quotient t/μ^2. The curve shown in Haldane's figure is for $t/\mu^2 = 0.01$. Replotted on Fig. 5.3 after transformation of the abscissa, it would fall very close to curve B, for which the value equivalent to t/μ^2 is $0.005/0.61^2 = 0.013$. The two sets of assumptions therefore provide very similar results. The shape of the computer curves can readily be modified to allow for progressive changes of one kind or another in selection, for change in migration rate from place to place, or for changes in dominance.

Disruptive selection

The term disruptive selection is used to describe selection favouring more than one mode or optimum in a population (Mather, 1953, 1955), that is, selection away from the mean in both directions (Thoday, 1960). It is synonymous with centripetal selection, a term used by G. G. Simpson (1953) and by Haldane. Characters studied in experiments on disruptive selection have usually been under multifactorial control (for example, in the experiments of Mather, Thoday, Robertson, Falconer and others), but we may think of the directions in which pressure is applied without making any assumptions about the genetic system involved. Mather (1955) and Clarke and Sheppard (1962) refer to selection for polymorphisms in Batesian mimicry as disruptive selection. Here, the phenotypes favoured are good mimics of one model or good mimics of another, and perhaps include the original non-mimetic phenotype itself, but intermediates between them are eliminated. A similar example is selection of unlike phenotypes in a prey species by predators that form a searching image (Clarke, 1964). Selection favours two or more optima which could be determined by major gene substitutions. Actually, there is evidence in the examples that have been studied (Clarke and Sheppard, 1960, 1962) that the switch from one form to another is actuated by a group of closely linked genes, the evolution of which is the result of the disruptive selection. But can disruptive selection in itself lead to the establishment of stable equilibria? When the system consists of two or more niches in an area where adults form a random mating unit, or when there are several sub-units with migration between them, the answer, discussed in this chapter, is that it can. If the population is completely panmictic, however, it cannot. In the simplest case of a pair of alleles at a single locus, disruptive selection is synonymous with heterozygote disadvantage and the equilibrium is unstable. It is also unstable if we select for the two homozygous extremes of a phenotype distribution under multifactorial control. When less extreme modes are selected genetic heterogeneity may be maintained, but then the modes are heterozygous combinations, and selection for them implies heterozygote advantage. In the case of a mimetic or an apostatic polymorphism the force which establishes the primary balance is frequency-dependent selection. The fact that selective pressure is disruptive may serve to modify the phenotype but does not account for the balance. If we can select for the two extremes of a multifactorial distribution and obtained rapid divergence, with reversion to the mean when selection is relaxed, then it may be asked why the genetic variability was available in the unselected line. The reason is probably that there is a good deal of heterosis at the loci controlling the character. A balanced situation following disruptive selection therefore also involves heterosis arising from gene effects other than those selected.

Having stated this, it must be remembered that attainment of complete panmixia is unlikely. If mating is random, but divergent selection is exerted on partially isolated groups of the ensuing progeny, a balance may be set up between selection, migration and perhaps segregation under a disruptive regime. The separate niches may be very distinct to the observer or barely distinguishable. In the latter case it would be possible for a stable polymorphism to be established in an apparently simple environment without evidence of heterozygote advantage or frequency-dependent selection. Indeed, evidence for heterosis would be hard to detect, for even if such selection occurred, samples taken without reference to niche and

scored for genotype would be likely to show an apparent deficiency
of heterozygotes if the equilibrium frequencies differed between
niches.

The balance between gene flow and divergence

The ability of gene flow to prevent local divergence has received a
good deal of attention in the literature. The equations discussed
indicate the conditions for balance in simple systems. Reduced to
their barest essentials they show that 1% selection in favour of a
gene in a population is balanced by approximately 1% immigration of individuals carrying its allele. Over all kinds of conditions the
intensity of the two pressures that will produce a balance has the
same order of magnitude. Some years ago, however, Mayr (1954)
suggested in an article that has stimulated much discussion that
migration may be a more efficient damper on genetic response than
had hitherto generally been imagined.

The view was criticized in a number of papers by Thoday and his
colleages (Gibson and Thoday, 1964, and earlier) dealing with studies
on populations of *Drosophila* in the laboratory. All the experiments
concern the response of lines of flies to selection for particular numbers of sternopleural chaetae. For example, Thoday and Boam (1959)
applied 50% gene flow between sub-populations selected for different chaeta numbers, and showed that nevertheless the sub-units
would respond to divergent selection. They argued that there is no
theoretical need for an isolation barrier between lines or races before
genetic divergence can occur, quoting Mayr for the view that such a
need exists. Now, in these experiments the variability in bristle number
is shown to be determined by several genetic factors which sometimes segregate as effective single units. During part of the experiment
the system was formally equivalent to a three-allele polymorphism,
only two of the three alleles being present in any one of the four
selected lines. When the procedure they employed is studied it is easy
to see that the results could accord with the simple alegbra, selection
being sufficiently intense to overcome or balance the migration
between selected lines. In a later paper (Gibson and Thoday, 1964)
a lower selective intensity was used. In each generation the 25%
fraction with the highest number of bristles and the quarter with
the lowest number were selected from a sample derived from four
lines, the other 50% being rejected. The selected individuals were
then mated assortatively (high by high and low by low) or disassorta-
tively (high by low and low by high) to produce four sets of parents
of the next generation. Divergence of the high line and the low line
rapidly took place. The intensity of selection in each line and the
amount of migration between them varied with degree of divergence

as the experiment proceeded, but even in this experiment, and certainly in the earlier ones, the pressures applied are at the extreme end of the range to be found under natural conditions. Fifty per cent of the sampled population is lost in each generation; on a simple genetic hypothesis the loss could involve lethality of unfavourable genes in the assortative lines. The selection is sufficiently extreme for divergence to be expected. Streams and Pimental (1961) also provide data on the balance between selection and migration in experimental populations of flies, which do not disagree with the simple mathematical relations.

The point emphasized by Mayr (1954, 1963) is that the effect of migration may be conditioned by the amount of interaction which occurs between loci. He suggested that the regular introduction of alien individuals leads to selection of indigenous factors which are tolerant of immigrant genes. Conversely, selection in isolation promotes rapid development of 'intolerant' combinations. Presumably tolerant genes are strongly buffered ones that have a constant effect, and therefore the same fitness, with whatever allele they are paired. They are likely to be top dominants in an allelic series, and to have fitnesses independent of frequencies at other loci. Intolerant gene pools are ones where fitnesses of genes at one locus are very sensitive to changes in gene frequency at other loci.

The tolerant gene pool hypothesis is the one to which simple mathematics can be applied and the balance between selection and migration is straightforward; but immigrants to populations where there is strong epistatic interaction between loci naturally have an effect that cannot be assessed by considering one locus alone. If the fitness of an allele at one locus is affected to a marked extent by presence or absence of particular alleles at another, then a few migrants could have an unexpectedly profound influence if only one locus is studied. Suppose we have a locus A_1A_2, the two alleles of which produce different enzymes. The heterozygous combination may have an advantage over both homozygotes in the presence of another gene B but be disadvantageous to A_1A_1, though not to A_2A_2, in the presence of b. The introduction of quite low frequencies of immigrants from a bb population would change the A locus from a balanced polymorphic state to one where A_2 was deleterious, so that even a low level of gene flow precipitates a shift from stability to extinction. In a system where stable and unstable equilibria exist, the addition of immigrants from outside may have the same effect, causing the gene frequency to move to the region where the polymorphism is out of balance.

A gene pool of interacting loci with a high degree of integration is an attractive hypothesis. Before the new balance is fully established a small amount of immigration may inhibit its development. After its development the different integrated gene pools (races or micro-races) may exist side by side without losing their genetic integrity. The possibility has been discussed in connection with variation in *Cepaea nemoralis* by Goodhart (1963) (for) and Cain and Currey (1963a, b) (against). This example is also discussed in an extensive paper on factor interaction by Wright (1965). There is, however, little positive experimental evidence that the required degree of integration can be established. The kind of data needed should arise from studies of systems such as the polygenic DDT resistance of *Drosophila* (King, 1955) in which a marked loss of fitness occurred in the F_2 after hybridization between selected resistant strains. In this case, however, the drop in the F_2 was made up by an unexplained recovery in the F_3. On the whole, the evidence for marked epistatic effects on fitness is lacking, whereas it should be readily obtainable if the hypothesis is generally applicable. Certainly the experiments of Thoday and his associates do not indicate the existence of integration.

NON-RANDOM MATING

So far all the selective processes discussed have been treated as if they were based on the random assortment of gametes, but there is no reason to suppose that mating must remain random. If a species is separated into races with a zone of contact between them preferential assortative mating by members of the same race takes place if the courtship signalling mechanisms have diverged sufficiently to reduce inter-racial recognition. When the hybrids are less viable than the progeny of within-race crosses selection favours an increasing degree of assortative behaviour. The ultimate result may then be specific divergence even when there was initially a certain amount of gene flow. This consequence, and habitat selection by animals from different niches, which may also involve modification of mating behaviour, are mentioned in the previous section. Assortative mating is certainly likely to play a part in the polymorphisms based on divergent selection and gene flow.

The reverse process of disassortative or negative assortative mating, where each genotype associates preferentially with an unlike genotype, will in itself tend to establish a polymorphism. The sexual systems of plants and animals are highly developed examples of this effect. Each genotype produces zygotes by fusion of its gametes with gametes derived exclusively from individuals of the alternative genotype. Genes at autosomal loci in a bisexual species may also be maintained at stable equilibria if their effect includes a tendency for individuals of one phenotype to mate preferentially with an unlike phenotype. Very permanent polymorphisms can result from such behaviour. One example is the series of self-incompatibility alleles found in several species of plants, but examples from animals are rare.

The tendency to disassortative mating is most easily observed when there is a distinct visual polymorphism. It might be anticipated among the diverse phenotypes of man in Europe, given the curiosity of human beings, but is not certainly known to occur. It has been looked for in vain in *Cepaea nemoralis* (Lamotte, 1951). The matings of that species of snail recorded by Lamotte under natural conditions were completely random with regard to the shell colour and pattern phenotypes of the pairs involved. Non-random mating between morphs occurs in many species of butterflies with mimetic and other polymorphisms, such as *Hypolimnas misippus* (Stride, 1956) and various other Batesian mimics (Brower, 1963), and in the non-mimetic fritillary *Argynnis paphia* (Magnus, 1963). In these examples, however, the polymorphism is limited to the females and the mating is preferentially assortative between the male and the typical, or male-like, female morph. This behaviour tends to eliminate the un-favoured morph unless there is a suitable selective counterbalance such as a mimetic advantage. Similarly, the white eyed mutant *w* in laboratory stocks of *Drosophila melanogaster* has a disadvantage at mating when present with the wild type, and the black coat colour in the mouse has a disadvantage compared to brown (Levine and Lascher, 1965). Mating behaviour is non-random, but the reciprocal behaviour that would be necessary to maintain a polymorphism does not exist. The most famous case of disassortative mating which would promote a balance is that discovered by P. M. Sheppard for the *bimacula* gene of the moth *Panaxia dominula*. There are three geno-types, the heterozygote *medionigra* differing in appearance both from the typical and from the mutant homozygote *bimacula*. Sheppard set up matings by placing together in a small enclosure one adult of one sex with two adults of the other sex having either the same geno-type or a different one from the first. Whether the sex represented by the single individual was male or female there was a tendency for mating to take place between the unlike pair (the most extensive data are in Sheppard and Cook, 1962). In a total of 199 trials there were just under 37% of like pairings. This appears to be due, in part, to rejection by females of males of like genotype. It is not known how behaviour is modified in the wild, where several males may be available to mate with any one female, but judging from the experi-mental results, the fact that males may mate several times while females do so only once, and the tendency of males to emerge as adults before females so that a situation of competition between males may readily arise, this behaviour should favour polymorphism. The gene frequency appears to have been tending to an equilibrium state in the only colony where, apart from introductions, the *medionigra* gene is known to occur (see Ford, 1964), and this behavioural trait is alone among the pleiotropic effects of the mutant gene in raising its fitness above that of the wild type.

Types of disassortative mating

The consequences of different types of mating system in terms of the way they change genotype and gene frequencies are not always obvious. To illustrate the general features we shall start from first principles and consider some different patterns of behaviour in a haploid organism with two mating types. The effect of genetic segregation in a diploid will then be added.

Suppose there are two genotypes A and a which start at frequencies l and m. On the basis of random mating we should have associations at the following frequencies.

genotype	A	a
A	l^2	lm
a	lm	m^2

Zygotes are produced, which then undergo reduction division to form the next haploid generation in these frequencies:

zygote	frequency	progeny	
		A	a
AA	l^2	l^2	—
Aa	$2lm$	lm	lm
aa	m^2	—	m^2
Total	1	l	m

The frequencies are unchanged by the round of mating, but several different patterns can be envisaged which cause gene frequency to be modified.

(a) A possible form of disassortative mating would require that individuals have one chance of mating during their lives, and do so with probability 1 if the pair is unlike but $(1 - g)$ if the pair is of like genotype. By the end of the round there will therefore be g encounters of like individuals which do not result in zygote formation. The modified pattern of association is as follows.

	A	a
A	$l^2(1 - g)$	lm
a	lm	$m^2(1 - g)$

zygote	frequency	progeny	
		A	a
AA	$l^2(1 - g)$	$l^2(1 - g)$	—
Aa	$2lm$	lm	lm
aa	$m^2(1 - g)$	—	$m^2(1 - g)$
Total	$1 - g(1 - 2lm)$	$l(1 - gl)$	$m(1 - gm)$

The new frequency of A is

$$l_1 = \frac{l(1 - gl)}{1 - g(1 - 2lm)}$$

The frequency will therefore be unchanged by the process of conjugation if

$$l = 1 - 2lm$$

or

$$2l^2 - 3l + 1 = 0$$

which is true if $l = 1$ or $l = \frac{1}{2}$.

To put this requirement another way we have

$$\Delta l = \frac{-glm(l - m)}{1 - g(1 - 2lm)}$$

and at the non-trivial equilibrium where $l = m = \frac{1}{2}$ we have

$$\frac{d\Delta l}{dl} = -\frac{g}{2 - g}$$

A tendency to favour unlike pairings (g positive) therefore results in a stable equilibrium at equal frequencies of the two mating types.

(b) It may happen in a species where unlike mating is preferred that each individual has several opportunities to pair. In this event the similar individuals which do not pair first time may do so later; and we will suppose that all members of the population eventually succeed. This means that individuals of whichever type is the rarer mate more than once. We have $A \times A$ conjugation at frequency $l^2(1 - g)$ and $a \times a$ at frequency $m^2(1 - g)$. The $g(l^2 + m^2)$ of possible like $\times$ like matings that are not achieved are divided proportionately between the $A \times a$ and $a \times A$ classes, so that the total in each category is $lm + \dfrac{lm \times g(l^2 + m^2)}{2lm}$ or $lm(1 - g) + \tfrac{1}{2}g$. It follows, in the same manner as before, that

$$l_1 = l(1 - g) + \tfrac{1}{2}g$$

or

$$\Delta l = -g(l - \tfrac{1}{2})$$

and

$$d\,\Delta l/dl = -g$$

There is therefore a stable equilibrium at $l = \tfrac{1}{2}$.

If $g = 1$, pairing occurs only between unlike types, and the equilibrium gene frequency is reached after one generation, a fact indicated by the value of -1 for the regression of Δl on l. When g is less than 1 there is incomplete outbreeding and it is doubtful if g should be constant. In these circumstances we can visualize the mating of like individuals proceeding at an unimpaired rate unless a choice is available, but being modified by the element of choice. The likelihood of a choice varies with frequency, so that deficiency of the two like $\times$ like classes should be frequency-dependent.

(c) To choose a simple frequency-dependent scheme, we may represent the $A \times A$ pairings as $l^2(1 - gm)$ and the $a \times a$ as $m^2(1 - gl)$. Adding the residue onto the unlike-pairing classes provides $lm(1 + \tfrac{1}{2}g)$ for each of them. The frequency in progeny is now

$$l_1 = l + glm(1 - l)$$

and

$$\Delta l = glm(\tfrac{1}{2} - l)$$

so that

$$\frac{d\Delta l}{dl} = 2g(1 - 6l + 6l^2)$$

$$= -\tfrac{1}{4}g \text{ at } l$$

(d) Another use of frequency-dependent mating coefficients has been made by Williamson (1960) in discussing the polymorphism of *Panaxia dominula*. Adapting his approach to the haploid model, we may assume as he did that each morph is twice as willing to conjugate with the other as with its own type. The $A \times A$ matings can be represented as $l^2\left(\dfrac{1}{l-2m}\right)$, and the $a \times a$ as $m^2\left(\dfrac{1}{2l+m}\right)$. The factors vary from $\frac{1}{2}$ when the kind is very rare to 1 when it is very common. For $A \times a$ pairings we have $\frac{1}{2}lm\left(\dfrac{2}{l+2m}\right) + \frac{1}{2}lm\left(\dfrac{2}{2l+m}\right)$, and likewise for the $a \times A$. Simplifying the expressions we get the following table.

	A	a
A	$\dfrac{l^2}{2-l}$	$\dfrac{3lm}{2+lm}$
a	$\dfrac{3lm}{2+lm}$	$\dfrac{m^2}{1+l}$

zygote	frequency	progeny	
		A	a
AA	$l^2/(2-l)$	$l^2/(2-l)$	—
Aa	$6lm/(2+lm)$	$3lm/(2+lm)$	$3lm/(2+lm)$
aa	$m^2/(1+l)$	—	$m^2/(1+l)$
Total	1	$\dfrac{l(6-7l+4l^2-l^3)}{(2-l)(2+lm)}$	$\dfrac{2-l^2-l^4}{(1+l)(2+lm)}$

Using the total for the A progeny it can be shown that

$$\Delta l = \frac{-lm(2l-1)}{2+lm}$$

so that a non-trivial equilibrium again exists at $l = \frac{1}{2}$, and

$$\frac{d\Delta l}{dl} = \frac{1-6l+6l^2}{2+lm}$$

$$= -\frac{2}{9}\text{ at }l$$

There is a close correspondence to the results of section (c).

These four systems of disassortative mating all have similar properties, providing the same equilibrium but different curves of Δl on l. The second of them is a straight line of negative slope such as arises from the action of migration or mutation, while the other three give curves akin to those for selection. Which one is most appropriate to an outbreeding pattern in a haploid organism depends on the biology of the species concerned. The different examples show that stability of the equilibrium follows from the fact of outbreeding and does not require that the behaviour be frequency-dependent, or that genetic segregation be involved.

Turning now to diploid species we may see the effect of segregation in the case of a completely dominant gene. Let the frequencies of AA, Aa and aa be d, $2h$ and r, while $p = d + h$ and $q = h + r$. The example comparable to case (a) above has the following mating matrix.

	$A-$		aa
$A-$	$d^2(1-g)$	$2dh(1-g)$	dr
	$2dh(1-g)$	$4h^2(1-g)$	$2hr$
aa	dr	$2hr$	$r^2(1-g)$

mating	frequency	progeny		
		AA	Aa	aa
$AA \times AA$	$d^2(1-g)$	$d^2(1-g)$	—	—
$AA \times Aa$	$4dh(1-g)$	$2dh(1-g)$	$2dh(1-g)$	—
$AA \times aa$	$2dr$	—	$2dr$	—
$Aa \times Aa$	$4h^2(1-g)$	$h^2(1-g)$	$2h^2(1-g)$	$h^2(1-g)$
$Aa \times aa$	$4hr$	—	$2hr$	$2hr$
$aa \times aa$	$r^2(1-g)$	—	—	$r^2(1-g)$
Total	$1 - g(d^2 + 4dh$ $+ 4h^2 + r^2)$	$(d+h)^2(1-g)$	$2(d+h)(h+r)$ $- 2gh(d+h)$	$(h+r)^2$ $-g(h^2+r^2)$

The terms in d, h and r could all be converted directly to terms in p and q if the frequencies were still in Hardy–Weinberg ratio. Although this is not the case it remains true that $d + h = p$, $h + r = q$ and $d + 2h + r = 1$. The new value for the frequency of A will therefore be found in terms of d, h and r as

$$p_1 = \frac{(d+h)^2(1-g) + (d+h)(h+r) - gh(d+h)}{1 - g(d+2h)^2 + r^2}$$

$$= \frac{p[1 - g(1 - r)]}{1 - g[(1 - r)^2 + r^2]}$$

From this equation we get

$$\Delta p = \frac{-gpr(1 - 2r)}{1 - g[(1 - r)^2 + r^2]}$$

Stationary points therefore exist at $p = 0$, $r = 0$ and $r = \frac{1}{2}$. None is affected by the value of g. If g is positive the numerator is positive for small p (when r is close to 1), and negative when there is large p, and r is a lot less than $\frac{1}{2}$. The non-trivial equilibrium is therefore stable.

Having shown that a balanced state exists when the population comprises half each of the dominant and recessive phenotypes, the genotypic frequencies may be found by putting the numerical values at equilibrium into the equation for the new frequency of aa. When this is done the equation

$$r_1 = \frac{(h + r)^2 - g(h^2 + r^2)}{1 - g[(d + 2h)^2 + r^2]}$$

becomes

$$\frac{1}{2} = \frac{h^2(1 - g) + h + \frac{1}{4}(1 - g)}{1 - \frac{1}{2}g}$$

so that

$$h = \frac{-1 + \sqrt{(2 - g)}}{2(1 - g)}$$

It is therefore possible to find values for genotype and gene frequencies at different values of g. On the de Finetti diagram they occupy the straight line $r = 0.5$ between the Hardy–Weinberg curve, when $g = 0$, and the side where $h = 0.5$ when $g = 1$. This is the line ab in Fig. 4.1.

Li (1955a) shows these properties to hold when the mating system corresponds to example (b). If $g = 1$ the system described is the sex polymorphism. All individuals mate with others of unlike type, and the final position where $d = 0$ and $2h = \frac{1}{2} = r$ is reached at once. This explains why a genetic polymorphism where one allele is dominant in effect results in the establishment of two genotypes at equal frequencies, although, as discussed in Chapter 4, arrival at a condition where $g = 1$ in so many outbreeding systems is presumably the result of a long period of evolution of the loci concerned. When the tendency to disassortative mating is incomplete a model in which the outbreeding factor is frequency-dependent is more appropriate here as in the haploid example.

6

POPULATION GROWTH AND LIMITATION

The processes studied in population genetics are those which change phenotype frequency and gene frequency. It is possible for the ratio of two types to change simply because one type is increasing in numbers more rapidly than the other, in which case the agency involved would be their different reproductive potentials; but very often there is selective elimination. The factor behind differential elimination is the restriction exerted on a population by its environment, because each individual is potentially capable of contributing more than one offspring to the next generation. Consequently, although selective coefficients can be, and have so far been, defined with no reference to numbers but only to relative frequencies, population dynamics and population genetics are often interrelated. For this reason the necessary elements of population dynamics are outlined in this chapter and some generalizations that will be useful later are introduced.

An early mathematical statement of geometrical increase was made by Linnaeus (see Skellam, 1955; Cole, 1957). He showed that if an annual plant produced only two seeds and if the offspring did the same in the next generation, and so on, then in twenty years the total population would exceed one million. The abundance resulting from this kind of increase was recognized very much earlier, and is embodied in the *Arabian Nights* story of the man who wanted as a reward one grain of wheat corresponding to the first square of a chessboard, plus two for the second, four for the third, eight for the fourth and so on for each square. This total is $\sum_{n=0}^{n=63} 2^n$, about 10^{19} grains. If each grain weighs $0\cdot1$ gm the total is more than 10^{11} tons. From the biological viewpoint, most populations with a potential reproductive capacity greater than unity must have become as large

Population growth and limitation

as they can be at some time, even if they do not currently exist at the saturation density.

The situation may be described formally by the equation

$$N_{n+1} = cN_n \tag{1}$$

where c, which may be any positive number and not necessarily an integer, is the net rate of increase, N_n is the number of individuals in the nth generation and time is measured in generations. In particular,

$$N_1 = N_0 c, \quad \text{and} \quad N_2 = N_1 c = N_0 c^2$$

so that for the nth generation

$$N_n = N_0 c^n \tag{2}$$

and

$$\log N_n = \log N_0 + n \log c \tag{3}$$

which is a straight line with slope $\log c$. This gradient defines the unrestricted rate of increase.

If Linnaeus had chosen protozoa reproducing by binary fission then the model would have described the situation exactly, every point lying on the line. But strictly speaking a plant cannot produce seeds after it dies so that there are two seeds plus the parent for a short while before the seeds have grown into the two plants of the second generation. The series is: 1, 3, 2, 6, 4, 12 ···. If all the plants in the species are immortal but reproduce only once the series would be 1, 3, 7, 15, 31 ···. Innumerable other possibilities are open. For example, an animal may reach maturity after a certain time interval, produce one pair of offspring, survive an equal interval, reproduce again and then die (see Cole, 1954). If the process is repeated by the progeny this series will begin 1, 3, 9, 24, 66, 180 ··· where each term from the fourth onwards is twice the sum of the two preceding ones.

Although the exact number of individuals present at one time may be difficult to determine, all such processes will converge on the line defined by (3) while growth proceeds in this way. The equation and its parameters are abstractions from the true situation but they describe the most important properties of growth. It will become increasingly obvious later that population processes may be described in ways which take account of every detail or in ways which summarize the general nature of the events taking place. Very often the latter are the most useful descriptions because they allow important generalizations to be made which are not unduly obscured by the details.

To make the present situation more general we may start by writing (2) as

E

$$N_t = N_0\, e^{rt} \tag{4}$$

where t is any time and $c = e^r$. This is the exponential growth equation, the exponent r being called the intrinsic rate of increase. When time is measured in units of one generation it is the natural logarithm of the net rate of increase. In general it is a more useful parameter than c because its value is directly proportional to the length of time indicated. For example, if the intrinsic rate is 6 for a ten-year period it is 3 per five years and 12 per twenty years, whereas simple proportionality does not hold for the net rate. In populations with overlapping generations, increase is expressed as the intrinsic rate usually for some arbitrary time interval such as per day or per year, rather than per generation, which is an interval that is often difficult to estimate.

Now the constant e is the sum to infinity of the series

$$1 + 1 + \frac{1}{2!} + \frac{1}{3!} + \cdots$$

and e^{rt} is the sum

$$1 + rt + \frac{r^2 t^2}{2!} + \frac{r^3 t^3}{3!} + \cdots.$$

We can therefore differentiate equation (4) with respect to t by finding the derivative of this series and multiplying by N_0. The result is

$$\frac{dN}{dt} = N_0\, e^{rt}\, r$$

$$= N_t r \tag{5}$$

In words this equation states that change in numbers with time is a constant function of number; and

$$\frac{dN}{N_t\, dt} = r \tag{6}$$

That is, change in number per individual per unit time (the specific growth rate) is a constant. The intrinsic rate of increase is composed of the input into the population of new births reduced by loss from mortality. We may make it explicit that these two processes operate by writing

$$r = m - d \tag{7}$$

where m is the instantaneous birth rate and d is the instantaneous death rate.

These equations describe what would happen if unlimited increase were allowed. In practice exponential increase can go on only for a restricted time, after which the population will tend towards the saturation level for the particular environment in which it is found. The reason is that when the space available is limited, competition for resources, in a broad sense, increases with numbers. When space is limited and constant, number and density become interchangeable terms, and we may speak of the maximum density that can be supported as the saturation density. Equation (5) must be replaced by one of the form

$$dN/dt = f(N)N \qquad (8)$$

The expression $f(N)$ indicates any decreasing function of N having some maximum value equivalent to r and falling to zero as the population nears the environmental limit. We may then replace $f(N)$ by $rf(N)$, to suggest that r is an intrinsic property of the organism reduced by competition or other effects brought on by the environment. For example, in man the average output per individual (the net rate of increase c) cannot be more than about 10, so that $r = 2\cdot3$, this being the maximum output that could be attained even under the most favourable conditions for increase. The relation between N and $rf(N)$ may be represented graphically (Fig. 6.1).

The numbers in the population (the abscissa) can vary between the limits zero, at which point $f(N) = 1$ and the potential rate of increase is r, to the saturation density K, at which point the rate of increase is zero. The wavy line between these points shows the general relation between potential increase and numbers. The simplest assumption would be that the two points were connected by a straight line, so that

$$rf(N) = r - bN$$

where b is the gradient; but in this case $b = r/K$, so that

$$f(N) = 1 - \frac{N}{K} \qquad (9)$$

We may therefore rewrite (8) as

$$dN/dt = rN(1 - N/K) \qquad (10)$$

and (4) as

$$N_t = N_0\, e^{r(1 - N/K)t} \qquad (11)$$

which are the differential and integral forms respectively of the logistic equation of Verhulst.

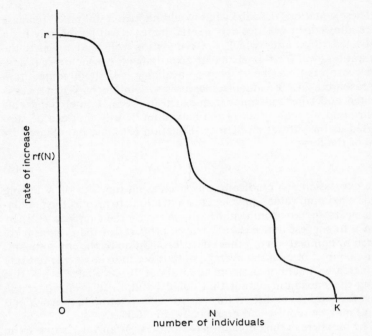

Fig. 6.1. The general relation which must exist between $rf(N)$, the
rate of increase in numbers, and numbers.

The logistic equation implies that the rate of increase is inversely
proportional to the amount of available space into which the population can expand. In practice the relation between $rf(N)$ and N is
likely to be more complex. Curves for experimental populations
(mostly of fish and insects) have been derived (e.g. Silliman and
Gutsell, 1958; Beverton and Holt, 1958) in some of which the relation is a curve falling steadily from the ordinate while in others the
rate of increase rises to a peak for a low value of N before falling off
to zero at K (Fig. 6.2), for it is quite possible that the reproductive
potential be severely impaired at very low densities just as it is at
high ones. Again it should be emphasized that the determination of a
suitable mathematical function describing the relation does not necessarily imply any biological relevance: a rather simple general relation
may arise by complex means. For example, in 1843 William Farr concluded that human mortality in English towns increased as the sixth
root of the population density (Cole, 1957). While this assumption
may have led to a very good fit to the data, the fact that the sixth
root is involved probably has no particular significance, and a better

fit might be obtained by deriving a model taking account of diffusion of pathogens, effiicency of transport of food, relative safety of different types of occupation and so on. Such a model would be more realistic but very much more complicated and laborious to determine. In the case of the logistic growth model a certain amount of biological reality is retained even though the underlying complex processes are ignored (see, for example, Watt, 1966, for a discussion of this problem).

The logistic model may be made to yield a further generality of population growth, the overshooting of the equilibrium level. As it is stated, the equation shows that the population tends asymptotically towards the upper limit K, because with each infinitesimal step forward in time the value of $rf(N)$ becomes slightly smaller. A value of N greater than K is impossible if N_0 is less than K. In a population with overlapping generations out of step with one another and with a relatively small potential to increase this must be true, but if the value of $f(N)$ is determined by population size an appreciable time

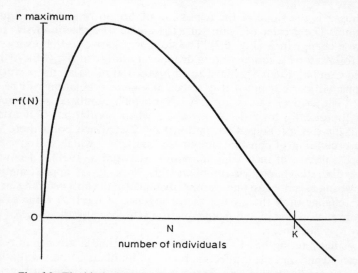

Fig. 6.2. The kind of relation between $rf(N)$ and N which probably occurs as a rule. At very low values of N the capacity for increase is lowered by the difficulty of finding mates, lack of beneficial social interaction, etc. At some high value of N the rate of increase declines to zero when the environment is saturated. It falls below zero if for some reason, such as a time-lag effect or immigration, a value of N larger than K is attained. Values for a curve of this kind are obtained empirically from the relation $rf(N) = \log_e \dfrac{N_{n+1}}{N_n}$.

before it becomes effective, then overshoot can occur. This fact may be demonstrated by writing the integral equation

$$N_{n+1} = N_n e^{r(1 - N_n/K)} \qquad (12)$$

In this form one time unit is taken to have elapsed, which may or may not be the generation time. Clearly if r is sufficiently large, N_{n+1} can have a value greater than K. The relationship can be shown with more precision by differentiating the difference equation $\Delta N = N_{n+1} - N_n$ with respect to N. Substituting from (12)

$$\Delta N = -N[1 - e^{r(1 - N/K)}] \qquad (13)$$

Now, all the N's refer to the same generation n, and

$$d\Delta N/dN = (1 - Nr/K)e^{r(1 - N/K)} - 1 \qquad (14)$$

At the equilibrium point where $N = K$, the right-hand side simplifies by substitution of 1 for N/K and we have

$$d\,\Delta N/dN = -r \qquad (15)$$

Thus, r is the slope of the regression of ΔN on N about the equilibrium. The greater the value of r the greater the reaction (ΔN) to a given change in N (Fig. 6.3). The slope is negative so that every rise is followed by a compensating drop, every fall by a rise (MacArthur and Connell, 1966; Maynard Smith, 1968a). If r is small the perturbations will die out but if it is sufficiently great oscillation will be set up. The critical values are $r < 1$, for which numbers will increase asymptotically towards K; $1 < r < 2$ when oscillation, if it arises, will die down (damped oscillation); $r = 2$ at which point there will be oscillation of constant amplitude; and $r > 2$ which will give rise to oscillation of increasing amplitude (Lewontin, 1958). Of course, for discrete changes in numbers this slope is an approximation, covering as it does an infinitesimal distance on the abscissa. The larger the change in N the greater the divergence of the true value of ΔN from the one estimated from equation (15), because the slope of ΔN on N is not linear.

A discrete lag has to occur between the time a given number is reached and the time it affects subsequent numbers before oscillations can be set up in this way. Many more elaborate studies of time-lag phenomena and oscillations have been made, including oscillations with a cycle length of several generations (Andrewartha and Birch, 1954; Nicholson, 1954; Hutchinson, 1948; Wangersky and Cunningham, 1957, etc.).

The equation discussed only allows oscillations with a cycle length of one generation, unless $r > 2$, when uneven fluctuations can arise. Regular fluctuations of greater length may be set up by interaction

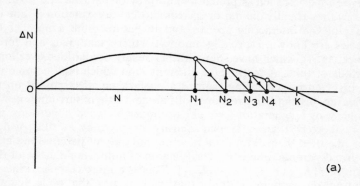

(a)

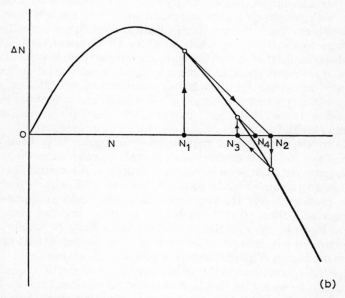

(b)

Fig. 6.3. ΔN plotted on N. $\Delta N = 0$ when $N = K$. The path of approach to K followed by successive values of N may be derived directly
from the graph, since $N_{n+1} = N_n + \Delta N$. Curve (a) shows the steady
approach to K when the slope of ΔN on N is less than -1, while in
(b) there is damped oscillation and the slope lies between -1 and
-2. For the relation of ΔN to N derived from the logistic argument
the slope of the curve about K is $-r$ approximately.

between species, such as predators and prey or parasite and host. In theory they usually die out or increase to cause extinction of one or other of the interacting species, but in nature quite a number of species are known, especially among northern mammals and birds (Lack, 1954), which have shown rather steady oscillations over long periods of time. A factor of a different kind which may commonly be involved in generating them is a phenotypic effect passed on to offspring, arising from response of parents to their surroundings. If progeny of a generation brought up under crowded conditions are less vigorous than a similar density of progeny of uncrowded parents, then their output is a function not only of parental but also of grand-parental numbers. The function of numbers might be of the form $K^{-1}[K - (1 - \gamma)N_n - \gamma N_{n-1}]$. This is a more complex kind of relation determining output, that will extend the cycle length. Another addition to the complexity arises from time-lag in populations with overlapping generations. In long-lived animals such as mammals and birds the age-specific birth and death rates vary from one age to another even when numbers are constant, and when numbers change the rates at different ages will not respond in identical ways. If such a population is started off below the saturation level, changes with time in the relative contributions of different age categories may lead to oscillation of some kind. The subject has been studied by Leslie (1948 and elsewhere) and others: it is reviewed by Moran (1962).

So far the processes of population maintenance have been described as if they were deterministic. In reality, even if something like the logistic equation is an adequate description, there must also be random factors involved. In addition, the level of K is likely to fluctuate randomly over the short term and may undergo systematic changes with time, so that numbers at a future time can never be determined exactly, and the population has a finite probability of extinction in any generation or point of time. Bartlett (1960) treats the mathematics of populations in a probabilistic manner.

Restricting ourselves to the simple deterministic system (equation 12) it will be seen that should K decrease or increase between one generation n and the next $n + 1$ a corresponding modification of numbers will ensue. If the new value is, say, half the previous value the population will drop to half its previous size provided that in generation n, $N = K$; but if for some reason N was very small, the change in K would have only a slight effect. Similarly, if the levelling off in numbers as the population approaches the steady state is achieved by increase in mortality (d) removing a progressively larger fraction of the unrestrained output (m), then the amount of mortality at any point is a function of the size of the population. Changes in numbers of these kinds in response to change in total environment may be said to be density-dependent. The term implies that a given amount of mortality or given change in a factor affecting the available ecological space has an effect at a later time on the population which depends on N. Usually mortality is greater or natality reduced when N increases, as considered here, but the opposite relation would also be possible. A second kind of relation can also occur, in which we can think of K as being a function of N. The most common situation of this kind must be interaction between animals and their food supply or between parasites and their hosts, which may result in oscillations in numbers of both the species involved. A factor behaving in this manner is called a delayed density-dependent factor by Varley (1958), a term which emphasizes the time-lag inherent in the system. The subject has been exhaustively treated by Nicholson (1954 and earlier), Volterra (in Chapman, 1931), Lack (1954) and many others. As an illustration, and restricting ourselves to the present elementary equation, we could write

$$N_{n+1} = N_n \exp r\left(1 - \frac{N_n}{K - bN_n}\right) \tag{16}$$

which, when $b = 0$, is equation (12). When b is not zero but positive, $(K - bN)$ decreases linearly with increase in N. It can now be shown that

$$d\Delta N/dN = \left[1 - \frac{rNK}{(K - bN)^2}\right] \exp r\left(1 - \frac{N}{K - bN}\right) - 1 \tag{17}$$

so that when $N = K$

$$d\Delta N/dN = \left[1 - \frac{r}{(1 - b)^2}\right] \exp -r\left(\frac{b}{1 - b}\right) - 1 \tag{18}$$

When $b = 0$ equation (18) is equal to $-r$ as before, but it is now possible for values greater than -1 to occur, even when r is less than 1, if there are positive values of b.

To complete the description of the population it is necessary to add another factor to take account of mortality the intensity of which bears no relation to density (due to density-independent factors). This may be done by writing

$$dN/dt = N[mf(N) - df(N)] - v \tag{19}$$

where v is a random variable with a mean of zero. The distinction between density-independent and density-dependent mortality is often a difficult one to make in practice. We might, for example, imagine a species of insect which pupates in holes in the ground that it is unable to make for itself. If there are a limited number of holes in a particular area they could constitute the principal density-limiting factor in the environment. Suppose that in one season half the holes are flooded so that only half the potential pupation space remains. If this catastrophe occurs before pupation the effect will be equivalent to reducing K by half and mortality will be dependent on larval numbers. On the other hand, if the flood occurs after pupation has taken place, about 50% of individuals will be killed irrespective of the number available (assuming random distribution of pupae and flooded holes), and mortality is thus described by a change in v. The distinction between density-dependent and density-independent factors, and the extent to which it is meaningful, is the centre of one of the most sustained controversies in ecology. It has some important consequences with respect to the genetics of populations, and for this reason will now be considered in more detail.

At an early date Nicholson (1933) distinguished between density-dependent mortality factors, which act to maintain a steady level in a species, and density-independent factors which tend to disturb it. Following development of the argument by various authors the density-dependent factors have sometimes taken on the status of population regulators, one or a few of which act for all species, that maintain the natural balance of a community; while density-independent mortality intrudes from time to time to upset the balance. This separation into two categories cannot usually be demonstrated in practical ecology, and was vigorously attacked by Andrewartha and Birch (1954). They pointed out that agents of mortality such as weather, which are often considered to be density-independent, must frequently act in a density-dependent manner (see, for example, Andrewartha, 1957), so that no satisfactory distinction can be drawn. Furthermore, the density of a population may be influenced by many factors interacting in a complex way with one another. Following Andrewartha and Birch, Browning (1962) has divided the external environment of individuals into five categories which are listed below. They are:

(a) Weather. Physical conditions such as temperature, humidity, sunlight, and in aquatic organisms, pH, salinity, etc., which may vary erratically but often have a diurnal, seasonal or some other cycle. Climate is a term for an average set of weather conditions in a particular region.

(b) Resources. Aspects of the environment that are used in some way and therefore influence survival or reproduction. They are principally related to food and living space, together with certain other special categories of space such as nesting sites, etc.

(c) Hazards. Inanimate objects which influence survival but are not used by the organism.

(d) Other members of the same species.

(e) Members of different species.

A general indication of the relations the various classes of factor bear to each other and their influence on capacity to increase is attempted in Fig. 6.4. The result should be a high degree of irregularity in fluctuation of numbers from one generation to another, but

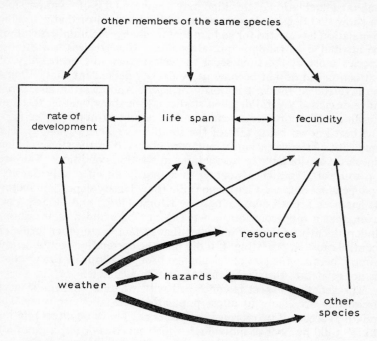

Fig. 6.4. Effect of the environment on capacity for increase. The general relationship illustrated by a graph of $rf(N)$ on N is in fact the result of many interacting factors. In this diagram the three boxes represent the components determining the capacity for increase of an individual. They are more or less interdependent. Below the boxes are four of the five categories of environmental effect proposed by Browning (1962), showing their most likely direct effects on capacity for increase and the probable interactions between them. Above is the fifth component: the other members of the same species. They exert a direct influence varying with density and each one of them is influenced by the components illustrated below. $A \rightarrow B$ means A has an effect on B. The thick arrows at the bottom indicate that the environmental effects act on each other.

given sufficient density-dependent control, a low probability of extinction. This is probably a fair description of the behaviour of population numbers in most animal species. Andrewartha and Birch and others have gone on, however, to stress the importance of factors acting in an effectively density-independent manner to determine numbers, with a corresponding neglect of the regulating role of density dependence. But populations in which there is no feed-back between output and numbers must ultimately increase unboundedly or fall to extinction. In certain circumstances, especially in marginal habitats, fluctuation of this kind may occur, and it is often impossible to show that density-dependent control is involved or that a specific population has existed for so long that its changes in numbers cannot be attributed to random mortality alone. Nevertheless, since most species most of the time seem to have a very low probability of extinction but do not become pests, density-dependent control must be involved in the vast majority of cases. Andrewartha and Birch have discussed several detailed studies which show weather to be of paramount importance in causing fluctuations in numbers, perhaps the best known being that of the small insect *Thrips imaginis*. The majority of the variation of numbers observed in the study of that species is attributable to fluctuations in weather conditions. They go on to assert that, since weather is acting in an effectively density-independent manner, there is no room for a density-dependent factor. It has been argued convincingly by Klomp (1962) and others, however, that a high correlation between numbers and a factor acting independently of density indicates that in fact some other feedback control must be operating. If it did not, the correlation would be low simply because the size of the change brought about by the weather is not related to the size of the population available.

The genetical aspect is concerned with whether changes in gene frequency are likely to affect population size and vice versa. The relative success of two forms in a species need have no direct relation to the total number of individuals which survives. Thus, a mutation may arise which increases the bearers' resistance to a density-independent factor, but if density-dependent effects appear later in the life of the individual, substitution of this gene for its allele will have no effect at all on population density. A change in gene frequency will then only lead to an increase in numbers if the character affected alters in response to a density-dependent factor (see Haldane, 1953). But of course it is not certain, in any specific instance, that a subsequent density-dependent factor will intervene. In addition, it is often impossible to decide into which of the two categories a factor falls: its proper designation may depend on an interaction between it and a second factor which has not been considered. To revert to the hypothetical example of the insect which pupates in holes in the ground, a gene which decreased the likelihood of a pupa dying in a flooded hole but did not affect the willingness of the larva to enter the hole would lead to an increase in density through decrease in v when flooding occurred after pupation—that is, when there is density-independent mortality—but would not alter the density if flooding occurred before pupation. Selection would therefore only occur if there were late flooding. If the gene also made the bearer more likely to enter a flooded hole when no other was available, selection for it would take place whether there was early or late flooding. The difficulty has been discussed by Williamson (1958).

Haldane (1956) has extended the argument by pointing out that under optimal conditions a species will exist rather close to the density-dependent limit, but in areas which are progressively less agreeable the density will increasingly become determined by density-independent factors. In consequence, a change in density following change in frequency of a gene affecting a density-independent factor should be more likely to be found in marginal habitats than in central ones.

7

AVERAGE EFFECTS OF SELECTION ON

POPULATIONS

Intensity

The term intensity is used in several related senses for situations where a general expression is required for the pressure to which a population is subject. Gause (1934) discussed the change in selection pressures as populations move from a state of exponential increase towards their saturation densities, defining the function i as the unrealized part of the potential increase expressed as a fraction of the realized part.

If a population is increasing in number exponentially the intensity of selective elimination is zero: it reaches a maximum as the population approaches equilibrium. If we write c for the realized increase at any density and C for the value attained with unlimited increase,

$$i = \frac{C - c}{c} \tag{1}$$

$$= C - 1$$

when the population is stable. Gause expressed the realized increase as a fraction of the possible increase. Calling this fraction n, we have

$$i = \frac{1 - n}{n}$$

If logistic growth is assumed we may then substitute for n the term $e^{r(1 - N/K)}/e^r = e^{-rN/K}$, so that

$$i = \frac{1 - e^{-rN/K}}{e^{-rN/K}}$$

$$= e^{rN/K} - 1 \tag{2}$$

which is identical to (1) when $N = K$. The value i varies from zero for very small N to $e^r - 1$ at saturation. Gause derived somewhat different expressions, one for the period of growth in numbers and one for the stationary phase, but they indicate a similar relationship. At equilibrium the intensity of selection is directly proportional to the maximum numbers of offspring that can be produced by a species. In the example given in Chapter 1 where the fitness of the ebony mutant of *Drosophila* changed with increasing density, the function of N ascribed to the ebony homozygote is an outcome of increasingly intense competition for survival—ebony flies are sensitive to the average stress imposed by competition. On the other hand, the fitness of the typical homozygotes compared to the heterozygotes does not vary with intensity of selection, and we can see from this example that a change in intensity does not have a necessary consequence in terms of the relative success of particular genotypes. When growth is exponential and two morphs have different intrinsic rates of increase their relative selective values differ but selective intensity is zero. Moreover, selective differentials are not necessarily greater in a stable population of a species with a high intrinsic rate of increase than in one with a low rate of increase, although the intensity of selection or of competition, as here defined, is greater. The reason is twofold. First, the selective differential under study may not arise from the action of a factor which influences numbers. Intensity in the present sense is only a useful concept when it does, and may be defined as the average selective pressure imposed by competition on phenotypes interacting with population controlling factors.

Secondly, even when the character studied does meet the requirement of density dependence, the outcome of increasingly intense competition depends on the differences both in the means and in the variances of fitness for the morphs considered, as Haldane (1930, 1932) has shown. The argument may be developed most easily by means of diagrams. Consider a continuously-varying character such as height or weight in two classes of organism. The distribution of this variable among adults in the two groups might be as shown in Fig. 7.1(a). Group A has a smaller mean value than group B, but their variances are approximately the same. A selective agent eliminates from both groups all individuals having a value less than x_i; those individuals which remain contributing proportionally to the next generation. The ratio of areas under the two curves is as $P : Q$ (where $P + Q = 1$); the fractions eliminated are α and β respectively. We may now write an equation to calculate a selective coefficient as in equation (1) of Chapter 1, viz.,

$$(1 - s)P : Q = (1 - \alpha)P : (1 - \beta)Q$$

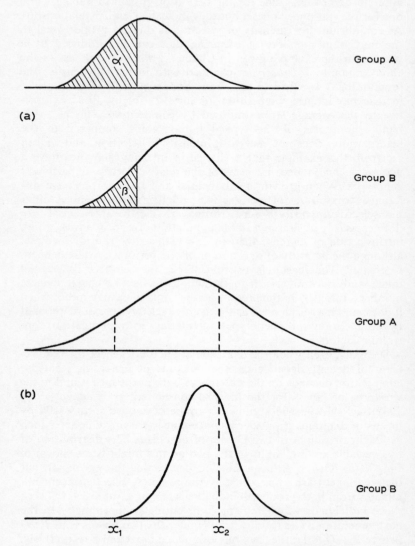

Fig. 7.1. Selective elimination from classes of individuals differing in means and variances of viability.

whence,

$$s = \frac{\alpha - \beta}{1 - \beta} \tag{3}$$

Now the intensity of competition as defined above is

$$i = \frac{\alpha P}{(1 - \alpha)} + \frac{\beta Q}{(1 - \beta)} \tag{4}$$

As i increases, increasingly greater fractions are eliminated from the left of the two distributions. Plotting s on i provides a curve similar to curve I in Fig. 7.2, the exact shape being determined by the form of

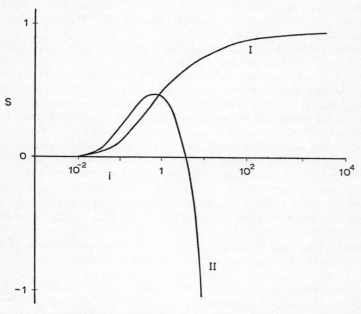

Fig. 7.2. Selective coefficients plotted on selective intensity for the two situations illustrated in Fig. 7.1. For explanation see pp. 145–6. From Haldane, 1930.

the two distributions and the distance apart of their means. Remembering that the abscissa in this figure has a logarithmic scale, the change in s with change in intensity of competition is relatively slow.

Now consider the two distributions shown in Fig. 7.1(b). Type A has a lower mean but greater variance than type B. We see that at the left side there is a region (up to x_1) where increasing elimination removes individuals of type A but does not affect those of type B, so that s progressively increases. Beyond this point both groups are affected and the slope of increase in s with i will gradually diminish until we come to a point (about x_2) where the majority of group B is eliminated but a comparatively large part of A still remains. The value of s becomes negative. The relation is shown in curve II, Fig. 7.2. A type which has a high mean fitness and small phenotypic variance is at an advantage over another type with lower mean fitness but larger phenotypic variance when the intensity is small, but at a disadvantage when competition is great. Intense competition favours variable response rather than high average response. 'Were this not so,' Haldane says, 'I expect that the world would be much duller than is actually the case.' A general treatment based on the normal distribution is given by Haldane (1930) together with a family of curves of this kind.

It may be asked why elimination from the two distributions should always take place from one extreme (from the left on the conventional kind of diagram). If the distributions are taken to represent any variable character there is no reason why elimination should be directional. It would be possible, for example, that the least successful individuals comprised the right-hand tail of A but the left-hand tail of B. In that case no simple relation between s and i could be derived. If variation in the character can be taken to reflect variation in viability, however, elimination will proceed as shown, the least viable individuals being removed. Intensity of selection is therefore a useful measurement so long as the character considered has a direct effect on ability to survive competition in the environment provided.

So far comparison has been made between two recognizably distinct categories, which may be races, strains or phenotypes. For example, they may differ in gene frequency at almost all loci, or at the other extreme they may be two groups of individuals homozygous for different alleles at a single locus but otherwise isogenic. If the variances differ in the latter case, one genotype provides better buffering than the other against environmental modification. Haldane (1954) has also considered the definition of selective intensity when there is stabilizing selection acting on a phenotypically varying character to reduce the variance from young to old stage within a single generation. The variation may be entirely environmental in origin, in which case selective elimination will not affect the variability in the subsequent generation, or it may derive from the combination of environmental effects and a multifactorial genetic system.

Haldane defines a measure of intensity as $S_0 - S$, where S_0 is the survival of the optimum type and S is the survival of the whole group. He quotes data of Karn and Penrose for survival of babies of

different birth weights. For females the survival rate of the optimal
class (7·5 to 8·5 pounds) over the first four weeks was 98·3%, while
the average for the whole sample was 95·9%. Hence $S_0 - S = 0·983$
$- 0·959 = 0·024$. In the terms considered above, if we take group A
to represent the optimum class, $S_0 = 1 - \alpha$ and $S = (1 - \alpha)P$
$+ (1 - \beta)Q$. It follows that $S_0 - S = - Q(\alpha - \beta)$.

To see the relation between this measurement and the selective
coefficients discussed in Chapter 1, imagine a very simple situation
in which the whole of group A consists of heterozygotes while group
B contains the two homozygous classes for a pair of alleles, so that
there is heterosis. If we now simplify by supposing 100% survival of
the optimum group we have $\alpha = 0$ and $S_0 - S = \beta Q$. Since the
selective coefficients for the two homozygotes may not be identical
βQ should be divided into its two components sp^2 and tq^2, so that

$$S_0 - S = sp^2 + tq^2 \qquad (5)$$

Now when the fitness of the heterozygotes is unity we can write for
the mean fitness (Chapter 1, equation 7 *et seq*),

$$\bar{w} = (1 - s)p^2 + 2pq + (1 - t)q^2$$
$$= 1 - sp^2 - tq^2 \qquad (6)$$

Consequently,

$$S_0 - S = 1 - \bar{w} \qquad (7)$$

The same relation may be derived for other genetic situations
(Spiess, 1962; Van Valen, 1965).

The intensity of selection is usually employed as a measure of
phenotypic selection however, the nature of the genetic basis of the
variability being unknown. If the variance of a character is reduced
between birth and adulthood, the extent to which the reduction is
carried to the next generation depends on the additive genetic contribution to the variance. The effective intensity is thus a function of
both $(S_0 - S)$ and the heritability. A change within the lifetime of
one generation results from selective elimination whether or not the
character eliminated is genetically determined. Selective intensity
may also be calculated between one generation and the next, but
this is likely to be misleading unless control of the character studied
is entirely genetic. If there is low heritability the index may merely
be measuring a change in environmental conditions.

In practice Haldane uses the value $I^H = \log_e(S_0/S)$, so that while
$(S_0 - S)$ has limits 0 and 1, I^H varies from 0 to $+\infty$. When the
intensity is low these two indices are approximately the same. Van
Valen (1965) employed the non-logarithmic measurement $I =$

$(S_0 - S)/S_0$, and this seems the preferable course since it gives a measure more directly related to the selective coefficients discussed before.

Selective intensities may be calculated from the change with age in the frequency distribution of individuals of different classes. The two curves in Fig. 7.3(a) represent the distribution of individuals in an

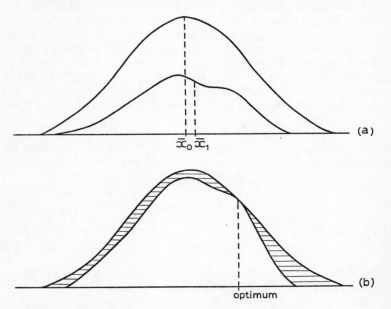

Fig. 7.3. Change in frequency distribution of individuals in a population following selective elimination. Modified from Van Valen, 1965.

ideal population before (upper) and after selection. In 3(b) the frequencies have been adjusted proportionally until the curves just touch each other. The value on the abscissa at which contact is made represents the optimum phenotype, which now has the same frequency before and after selection. The index I is the fraction of the total area which lies between the two curves, i.e. the proportion of what was available before selection which is lost by selection. I^H is the negative logarithm of the area below the lower curve. Haldane showed for variables distributed normally before and after selection that when the base population has a mean value $\bar{x}_0$ and standard deviation sd_0, and the survivors have mean and standard deviation $\bar{x}_1$ and sd_1, then

$$I^H = \log_e\left(\frac{sd_1}{sd_2}\right) - \frac{(\Delta\bar{x})^2}{2\Delta V} \qquad (8)$$

where $\Delta\bar{x}$ is change in mean $(\bar{x}_1 - \bar{x}_0)$ and ΔV is change in variance $(sd_1^2 - sd_0^2)$. We may also obtain from this equation the value of I because

$$I = 1 - e^{-I^H} \qquad (9)$$

so long as S_0 is set equal to unity (Van Valen, 1965). The following data taken from Inger (1943) for ventral scale number in males of the snake *Thamnophis radix* illustrate the use of these equations.

	mean	standard deviation
juvenile	154·6	3·63
adult	156·7	3·17

With the distributions for juveniles and adults considered to be normal, and the juveniles taken to represent accurately the distribution in the juveniles from which the adult sample was selected, we obtain $I^H = 0.136$ if the second term is ignored, or $I^H = 0.838$ when it is included. This is equivalent to $I = 0.127$ and $I = 0.567$, so that the intensity lies between 13% and 57%. The quite small but significant increase in the mean value makes a large difference to the estimation of intensity. Van Valen has provided formulae for estimation of intensity when there is truncation of one or both tails of the distribution and with reduction of the central part so that the variance increases. He has applied them to data from a fossil population, and shows that the estimates vary widely depending upon the assumptions about distribution which are made.

O'Donald (1968) commented upon the importance of making the correct assumptions about the frequency distributions before and after selection. To be realistic it is necessary to relate the variable x to the survival value. As a starting point he assumes that survival value decreases in each direction as the square of the deviation of x from its optimum value. He then derives equations from which intensity can be estimated, that take account of changes in x with respect to its mean, variance, skewness and kurtosis, between pre- and post-selectional stages. O'Donald obtains intensities of 9·5% and 8·6% for some data discussed by Van Valen and Weiss (1966) whereas they estimated 4·1 and 3·5 respectively, regarding the reduction in variance as due to truncation of the tails of the distribution. The estimations of I obtained from equation (8) provide 11·2% and 9·9%.

Figure 7.4 shows the results of an experiment on over-winter survival in *Cepaea nemoralis*. A sample of animals was left without food from the end of September to mid-April of the following year in conditions that were slightly warmer than those they normally experience during hibernation. In this time 44% of the sample died. The figure shows the size distribution of the original sample (upper curve) and of the survivors (lower curve). The % survival, or fitness, of each size class is given below, suggesting that the optimum diameter is 26 mm or greater. In the 26 and 27 mm intervals the average survival is 68%. When the lower curve is increased by 1/0·68 the

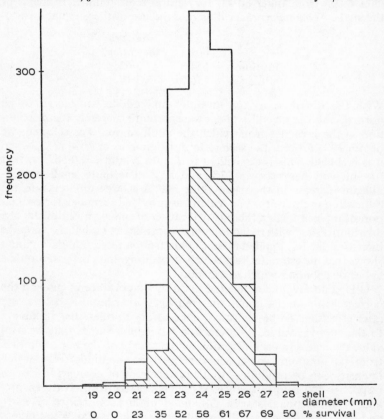

Fig. 7.4. Frequency plotted against maximum diameter of shell for a sample of the snail *Cepaea nemoralis* kept without food over winter. Upper curve – total sample. Lower curve – survivors. Below each size interval on the abscissa is shown the per cent survival within that interval.

fraction of the total area lying between the two histograms is 16·8%, which is therefore an estimate of selective intensity. Using Haldane's method the intensity is 15·2%, and by O'Donald's general method it is 13·6%.

In view of the sensitivity of the estimates to change in the initial assumptions they can give us no more than a general idea of the order of magnitude of selection. They also suffer from the defect that the optimum with which the average survival is compared must be chosen by the investigator. For example, the true optimum under the conditions of Fig. 7.4 may be a mean diameter greater than 28 mm. If such a value were used the estimated intensity would differ from the ones so far obtained. When we restrict ourselves to values within the range exhibited by the sample the choice of 25, 26, 27 or 28 mm as the optimum class is influenced by the sampling error in survival within each class. Because numbers in each class are comparatively small the sampling error may be great, and, since it affects the decision as to which class is optimal, this error will always lead to over-estimation of the intensity if the direct method of calculation is used. It is therefore best to fit continuous distributions to the data in order to find the intensity. I am grateful to Dr O'Donald for pointing this out to me.

Another value referred to as intensity of selection, and called i, is a measure of the force exerted in artificial selection programmes to shift the population mean. It is the difference between the average value for selected parents and the average for the population as a whole (Falconer, 1960; Lerner, 1958), measured in units of standard deviation. If the base population has standard deviation sd and mean $\bar{x}_0$, and the parents of the next generation are taken from one end of the range so that their mean value is $\bar{x}_1$, then

$$i = -\frac{\Delta\bar{x}}{sd} \tag{10}$$

When the base population is more or less normal in distribution and the rejected individuals are removed by truncation at a point x_i it can be shown (Falconer, 1960) that $i = z/p$ where p is the fraction selected and z is the height of the ordinate at the point of truncation.*

---

* The equation of the normal curve is

$$y = \frac{1}{\sigma\sqrt{(2\pi)}}\exp -\frac{(x-\mu)^2}{2\sigma^2}$$

When $\mu = 0$ and $\sigma = 1$, i.e. the distribution is standardized, this simplifies to

$$y = \frac{1}{\sqrt{(2\pi)}}e^{-\frac{1}{2}x^2}$$

By convention the simplified expression is called z and tables of z for different deviations may be found in Fisher and Yates (1963).

The value i can therefore be found, knowing either the shift in mean or the fraction selected. This index is used in animal and plant breeding as a measure of the expected response to directional selection.

Van Valen (1965) finds for the situation where there is truncation at one end of a normal distribution with zero mean and unit variance,

$$n = \frac{e^{-1/2x^2}}{(I-1)\sqrt{(2\pi)}} \tag{11}$$

where n is the new mean and x is the point of truncation. The mean n may in fact be positive or negative, the original mean being zero, but since we require a positive value of I adjustment may be made to the sign on the right-hand side. This expression may then be rewritten

$$n = z/(1-I), n \text{ positive}$$

or

$$n = -z/(1-I), n \text{ negative} \tag{12}$$

z being the ordinate of the standardized normal distribution as before. In order to deal with normal distributions with different standard deviations and means we have to find $n = (\bar{x}_0 - \bar{x}_1)/sd$. Therefore, from (10)

$$i = z/(1-I) = z/p \tag{13}$$

and

$$p = 1 - I$$

Consequently Fig. 11.3 in Falconer (1960), which shows the relation between i and p, is a graph of i on I if the scale of the abscissa is reversed, and is the same as Van Valen's (1965) Fig. 2.

To summarize: three measurements called selective intensity are distinguishable. The first (i) is the unrealized output of a population expressed as a fraction of the realized output. It is zero when there is unrestricted increase and reaches a maximum in a steady population, that is one less than the maximum net rate of increase for the species. If for any reason i exceeds this value the population must decrease in size. Since i increases with increase in intraspecific competition for resources it is a measure of density-dependent mortality. Change in gene frequency at one locus will not usually affect i, but if it is possible to measure the effect of differences in selective value in terms of i then a relationship between the genetics and dynamics of the population is being expressed.

The second index (I) measures the number of individuals which fail to survive because they do not possess the optimum phenotype, expressed as a fraction of those available to survive. In other words, it is the average mortality due to phenotypic diversity compared with the minimum mortality attainable if all individuals had the optimum phenotype.

The third value (like the first, called i) is used in plant and animal breeding terms to express the pressure to which a population is subjected in an artificial breeding programme. Both in this case and in the second the extent to which the pressure applied leads to change in the progeny generation depends on the heritability of the phenotypic characters selected.

Genetic load

The selective intensity I is concerned with the relative survival of optimal and suboptimal phenotypes. In some circumstances the relative survival rates of the different classes are identical with the relative fitnesses, w, of genotypes, so that $I = \dfrac{S_0 - S}{S_0} = \dfrac{w_0 - \bar{w}}{w_0}$, where w_0 is the fitness of the optimum genotype. When w_0 is considered to be the mean fitness of a population entirely composed of the optimum types, this index is known as the genetic load of the population.

As an example of its application consider the sickle-cell polymorphism in man. There are three genotypes, the typical homozygote AA, the heterozygote AS and the sickle cell homozygote SS. They have fitnesses in parts of West Africa of approximately 0·8, 1·0 and 0·5 (Allison, 1955). The polymorphism is stable and the fitness values indicate that of every generation at birth only 80% of AA individuals and only 50% of SS individuals reach reproductive age compared with a standardized 100% of the heterozygotes. If the population is at equilibrium the frequency of the A gene is $p = \dfrac{0·5}{0·7} = 0·714$, so that 20% of p^2, or 0·102, of the population and 50% of q^2, or 0·041, of the population die from causes associated with their genotype. Of course, some AS individuals may succumb either to malaria or to some effect of deficient haemoglobin, but heterozygotes have the maximum relative fitness (indicated by $w = 1$). A fraction $(s_1 p^2 + s_2 q^2)$, or 14·3%, therefore dies which would not do so if all individuals possessed the optimum genotype. This is the segregational genetic load relating to the sickle-cell condition, the penalty attendant on establishment of the genetic polymorphism.

The mean fitness of a population may be reduced below the optimal level by two other forces—the action of recurrent mutation continually introducing deleterious mutants and the action of migration, which has a depressing effect if the migrants are adapted to different conditions. These aspects are known respectively as the mutational and the migrational load. Each process introduces unfavourable variants, and the average depression of population fitness depends on the balance set up between rate of introduction and rate of elimination. Mechanisms tending to decrease the mutation rate and the willingness to interbreed with outsiders will be selected for, at least to the point where the drawback of load is balanced by the advantage of genetic flexibility, if such a point exists (Kimura, 1960, see ch. VIII). The concept of a segregational load, on the other hand, presents more problems.

One of the difficulties is that it is not clear what value to assign to w_0. In the present example it would be impossible for the whole population to consist of heterozygotes, so that the load is the measure of the extent to which the population falls short of an unattainable goal. Historically, the sickle-cell gene is a mutant which arose and was favoured in an environment where malaria caused deaths in a population almost entirely composed of AA individuals. When the S gene had just become established in the population, the mean fitness was 0·8, from which point it rose to 0·86 as the frequency moved to the equilibrium. In terms of genetic deaths after establishment of the polymorphism compared with those experienced when the mutant first arose, there has been a net improvement. The genetic load conferred by a polymorphism can only be considered to be a burden if the existence of the optimal heterozygous genotype implies the possible existence in the future of a gene having the same fitness when homozygous. Probably the only circumstance under which this is a realistic assumption occurs when the two alleles producing the high fitness become combined on one chromosome by duplication, following unequal crossing over. If they then retain their advantageous properties, the goal has been reached. Such a change is unlikely to occur.

The establishment of a polymorphism by heterozygote advantage depends on the relative proportions of the genotypes in each generation, but does not necessarily imply selective mortality of individuals. The effect would be the same if the offspring of different genotypes were produced in different amounts. There would then be unequal representation of the genotypes among the zygotes forming the next generation, but no more mortality than occurs in a monomorphic population. There is still a genetic load but it no longer involves inefficiency of use of resources or, in a human population, distress to the survivors.

Nevertheless, there is a further difficulty resulting from the genetic load, which may be illustrated by reference to another well-known example of polymorphism. Melanic forms of the moth *Biston betularia* first began to be favoured when industrialization began to blacken their environment with soot deposits. The number of deaths of adults resulting from predation increased by a large amount as the typical morph ceased to be cryptic. This fraction of mortality subsequently went down progressively as melanics became more numerous, until in areas with almost one hundred per cent melanics it attained a new steady level. The final mortality rate may be higher or lower than the original level depending on the relative crypsis of typicals on unpolluted and *carbonaria* on polluted backgrounds. The mortality accompanying the substitution of the *carbonaria* for the typical gene is the cost, or substitutional load, borne by the population during the process of change. The example of *Biston betularia* is used by Haldane (1957a) in a paper where he quantifies the limits imposed on selection by the number of deaths per generation that are required to bring about a gene substitution. He concludes that the number of deaths needed for the replacement of one allele by another is independent of selective intensity over a wide range of selective pressure, and is something like 30 times the number of individuals present in a single generation. When selection is intense elimination of all but the favoured individuals results in very heavy mortality, and the change takes place over a short time interval. At low selective pressures there are few deaths per generation but many generations involved, and the total number of deaths may in fact be greater. Within a single generation the reproductive capacity of a species sets the upper limit to the intensity of selection that can be sustained at any one locus, and, for a given average level of selection pressure, on the number of loci at which selection can act simultaneously. If f is the average fraction surviving deaths resulting from selection on one locus, then f^n is the survival when n loci are subject to selection together. It is therefore necessary that if_n be greater than 1, where i is the value of the selective intensity index reached in a stable population. This puts an upper limit on the number of loci which can be selected. The argument applies whether selection acts through mortality or output, because even in the latter case, selection implies reduced participation by some genotypes and increased participation by others in the formation of the succeeding generation, so that when there is heavy selection the burden of maintaining numbers falls on a small fraction of the population.

Most species have very high potential rates of increase. Study of their ecologies does not usually suggest that selection is anything like severe enough for populations to run the risk of extermination in this way. This is necessary to the view stated in Chapter 8 and elsewhere that most genes can be thought of as having effects independent of population density. Nevertheless, there are situations where the pressures do appear to come dangerously close to the limit. Experiments on *Biston betularia* show that during the transitional stage the majority of the population may have been eliminated in each generation. Suppose 75 % of adults are removed before contributing to the next generation and that selection at the same level acts on a locus that governs larval colour. Only 6 % of the original population present as eggs in each generation will survive to lay eggs in their turn. The output of *betularia* would sustain selective mortality at this level, provided the density-dependent checks act in the right places, but little room is left for selection on a third locus.

When the *carbonaria* gene was first observed, heterozygotes could be distinguished in appearance from homozygotes. Now there is complete dominance. One way for modification of expression to take place is for favoured pleiotropic effects to become dominant while unfavourable ones become recessive, so that, as Sheppard has pointed out, potentially transient polymorphisms become permanent. No populations are known composed entirely of *carbonaria* individuals, which suggests that balanced polymorphism has been established (see Sheppard, 1958). If the change in expression is due to modifiers at other loci, then simultaneous selection of at least two loci has been involved. If it is due to substitution of one kind of melanic allele for another, which is more likely considering the speed of the change, then only one locus is selected, but the result in either case is a balanced polymorphism with attendant segregational load. The change in environment has therefore led not only to high levels of mortality during the period of transition from a low to a high frequency of the melanic gene, but also to a slight addition to the segregational load of the species.

The conclusions reached by Haldane with respect to the cost of gene substitution apply equally to segregational load. If a population is thought of as being in a balanced state with selection acting only on polymorphic loci at equilibrium, the maximum reproductive rate sets a limit on the number of loci at which selective elimination can act independently at one time. In fact, populations are continually changing under fluctuating environmental pressures, so that there will be in addition a substitutional load component, and if transient polymorphisms tend to be converted to permanent ones the segregational load tends to increase. By studying a random sample of loci in wild populations of *Drosophila pseudoobscura* Lewontin and Hubby (1966) estimated that about a third of the total genome, or some 2000 loci, is polymorphic in nature. Similar fractions have been found by other workers. If all the polymorphism is maintained by heterosis the value f^{2000} has to be contained within the potential maximum net rate of increase of the species. Suppose that under ideal conditions each female is capable of laying 500 eggs. This requires that the value of f be 0·997 or more, so that if equilibrium frequencies are distributed about the mid-point, the average fitness of homozygotes at each locus, compared with a fitness of 1 for the heterozygotes, is 0·994 or more. The selective coefficients have an average value of no more than 0·6%. For populations of apparently vigorous flies which suffer a fair amount of random non-selective mortality, the mean coefficient must be much lower and lies in the same region as the average mutation rate. If it is higher, polymorphism must be much reduced. Yet heavy selection is readily detected in the wild and populations are polymorphic. Apart from the kind of evidence discussed in this and earlier chapters, the proportion of chromosomes carrying recessive lethals and semi-lethals in wild *Drosophila* populations is often 30% or more (e.g. Band and Ives, 1963; and see the extensive review by Crumpacker, 1967).

There are three possibilities to be considered which may reconcile these observations. They are as follows.

(a) The hypothesis about the action of selection is correct but the conclusion drawn is false. The argument as it stands indicates a vast falling short of the optimum when there is much polymorphism in a population, and implies an unattainably high potential reproductive capacity. But the optimum in question in the *Drosophila* example is represented by individual heterozygous for two thousand loci. Such animals have no more than a 10^{-600} chance of occurring. When the mean fitness of the population is compared with the fitness of a class of multiple heterozygotes present at a finite, though small, frequency in each generation the difference is very much smaller and quite realistic. This point was made by Sved, Reed and Bodmer (1967) when they showed that even if many loci are involved the variance in selective values between individuals is low, the extreme multiple heterozygotes being so rare as to play an insignificant part in determining it.

(b) The hypothesis is incorrect. These authors and several others (e.g. King, 1967; Milkman, 1967; Maynard Smith, 1968c) point out that the model for selection upon which the argument is based is not necessarily appropriate. It is not necessary, and it may not even be likely, that selection on the different loci is independent and multiplicative in effect. Instead, when individuals are normally distributed in fitness, selection may pick out the fitter ones with relatively high frequencies of heterozygous loci and eliminate the tail containing predominantly homozygous individuals. For a given level of mortality, the results of threshold selection of this type is to remove homozygous loci at a rate much higher than would be possible on the multiplicative model, with no reduction in fitness of the survivors. Analysis of specific situations shows that the number of polymorphic loci that can be maintained may sometimes be several orders of magnitude higher on the second hypothesis than on the first. The two models apply to different ecological conditions. The first one is applicable when loci govern the probability of death per unit time from causes directly ascribable to genotype, and when their action is independent. The second refers to situations where the contribution of different genotypes to the next generation depends on the outcome of intraspecific competition, and, whether or not competition is involved, to the joint action of several genes upon the same character. Melanism in larval and adult *B. betularia* would be an example of the first type, because selection of the second stage follows and acts upon survivors of the first. Stabilizing selection on a multi-factorially controlled character is an example of the second. In nature the two patterns of selection merge into one another. A genetically controlled disability may to a greater or lesser extent reduce success in all competitive spheres of life. It is certain, however, that the first hypothesis of independent multiplicative selection does not always apply, so that the problem of genetic load arising from multiple heterozygosis is much less severe than is predicted by it.

(c) The conclusion is correct and most polymorphisms are not maintained by heterosis with appreciable selection pressure. At the practical level the argument of Lewontin and Hubby should lead us to look closely at the evidence for the average size of selective coefficients at polymorphic loci. The smallest chemical changes in genes—the substitution of single bases—must often result in selectively neutral mutant isoalleles, especially where the change is to a synonymous triplet. It seems probable that selective values are unimodally distributed about the neutral point, the size of deviation depending more or less on the amount of chemical difference between the two alleles being compared. Experimental field studies have revealed very large selective differentials, but the scale at which an experiment or census can be applied to a population is necessarily limited, so that small selective differences are undetectable by direct means. The records therefore refer only to the tails of the distribution, and although they are very important in showing that the range of selective pressure is far greater than was first suspected, they do not, by themselves, indicate the variance. When sufficient observations are available, study of the shape of the observed part of the frequency distribution may throw light on the form of the complete curve. Meanwhile, the evidence that the expression of major genes varies with their genetic background, and that dominance may readily be modified by selection, suggests that many loci are polymorphic but have primary effects subject to very low selection. They are therefore on call to act as modifiers. This implies that polymorphisms subject to selection at a similar level to mutation pressures may indeed be quite common, so that the average coefficient has a value nearer to 1% than to the 10% indicated by studies of the primary effects of polymorphic loci. In summary, it seems probable that a high proportion of heterotic loci with large coefficients can be retained, since selection does not have the multiplicative property assumed by the first model, but study of polymorphisms subject to strong selection indicates the existence of many others on which selective pressures are slight.

8

ASPECTS OF GENETICS AND ECOLOGY

Mean fitness

In Chapter 3 it was noticed that mean fitness in a simple polymorphism always tends to maximize and reaches its maximum at equilibrium. This fact was used simply to find the equilibrium frequency, but the idea of change in mean fitness under selection has the status of a general principle. When selective values are constant a population will come to equilibrium at the point where the largest fraction of individuals has the highest attainable fitness. The fitnesses are defined relative to one another, and in practice they are measured after the event. A type which leaves a relatively greater fraction of progeny than another is described as the fittest. The tendency for mean fitness to maximize therefore means that those individuals establish themselves in a population which in the short term are best able to do so. The effect on gene frequency depends on the genotypes of the individuals in the classes grouped according to relative fitness. In a simple polymorphism with constant selective values, the mean fitness is $p^2w_1 + 2pqw_2 + q^2w_3$, or $\Sigma f_i w_i$, or simply $\overline{w}$. The largest fraction has the greatest possible fitness when $\overline{w}$ is at a maximum.

Wright (1949, and see Li, 1955a) gives the general change in gene frequency as

$$\Delta q = \frac{q(1-q)}{k\overline{w}} \frac{d\overline{w}}{dq} \tag{1}$$

The factor k is 1 for a haploid system, 2 for a diploid, and so on for higher levels of ploidy. For sex-linked genes the expression is approximately true if $k = 2/3$ and the mean fitness is the geometric mean of the $\overline{w}$ values of male and female. This equation is equivalent to equation (8) of Chapter 1, except that it is written in terms emphasizing $\overline{w}$. In this form it is easy to see that $\Delta q = 0$ when $d\overline{w}/dq = 0$.

Fisher's (1930) *Fundamental Theorem of Natural Selection* states that 'the rate of increase in fitness of any organism at any time is equal to its genetic variance of fitness at that time'. His demonstration of the equality is in terms of overlapping generations with no dominance. It may readily be seen to be true if we restrict ourselves to selection taking place between birth and maturity in one generation. Suppose the population starts with the frequencies p^2, $2pq$ and q^2, and relative fitnesses w_1, w_2 and w_3. After selection the remaining survivors will end up at frequencies p^2w_1, $2pqw_2$ and q^2w_3. The argument is then simplified if the w_i are arranged so that $\bar{w} = 1$. The final mean fitness is therefore

$$\bar{w}' = p^2w_1^2 + 2pqw_2^2 + q^2w_3^2$$

and the change in fitness brought about by selection is the difference between this expression and the first. The variance of fitness at the start is the sum of squares of deviations from the mean fitness, i.e. $\text{Var}_w = p^2w_1^2 + 2pqw_2^2 + q^2w_3^2 - \bar{w}^2$. But this expression is $\bar{w}' - \bar{w}^*$, which by definition is equal to $\Delta\bar{w}$. Consequently, $\Delta\bar{w} = \text{Var}_w$. This derivation, given by Li, leads to the same result as Fisher's if there is no dominance, but Var_w is greater than Fisher's value if there is dominance of fitness.

It is not nearly so easy to see that the relation holds from one generation to the next when random mating intervenes; and indeed there are situations where it has been shown not to be exactly true. Turner (1967) gives a good review of the subject. Fisher proposed that in general the rate of increase should be equal to the additive genetic variance. We can see in an intuitive way that the additive variance is the important feature by considering the relation between response to selection and the heritability of a character. Heritability is the additive component expressed as a fraction of the total phenotypic variance, and is equal to the regression coefficient of progeny mean values on the means of their parents (Falconer, 1960). The shift in mean in a group of progeny following selection is therefore equal to the difference in mean between the selected parents and the unselected population from which they are drawn, multiplied by the heritability. That factor is therefore a direct measure of selectability. Under natural selection it must be directly related to the improvement in fitness that takes place.

It is therefore at first sight disturbing to find systems where fitness does not change quite in this way, and even ones where it does not maximize. Some are trivial, such as the decline in fitness following a rapid change in the environment. Others result from the complexity of the genetic system. The mean fitness of a frequency-

* $\bar{w}^2 = \bar{w} = 1$.

F

dependent system does not behave in a simple manner (see Chapter 4), but a function derived from it, which has a common-sense meaning in terms of average fitness, does so (Li, 1955b). Similarly, Li has shown that when there is differential selection in different niches, $\bar{w}$ may be defined in such a way as to maximize at stable equilibria (see Chapter 5).

These examples indicate the main source of the difficulty. The general principle that nothing succeeds like short-term success can usually be demonstrated in terms of $\bar{w}$. When this is not possible it is the choice of $\bar{w}$ as indicator that is at fault. The curve of $\bar{w}$ on q sometimes embodies too little or the wrong kind of information for the purpose. Apart from the examples mentioned, this failure can occur when there is locus interaction (Moran, 1964) and with respect to the effect of the mating system and inbreeding (Turner, 1967; Li, 1967). Nevertheless, when fitnesses are defined in relative terms the mean fitness usually tends to increase. The consequences of the tendency will now be considered first with respect to the relation between relative fitness and output, and secondly, to selection for longer-term gains such as stability of numbers and storage of genetic variability.

Relation of relative fitnesses to output

It has been emphasized that the standard treatment of selection tells us nothing about the density or the rate of change in numbers of a population. Most of the time the study of selection cannot do so. If the factors controlling density are not affected by a difference in phenotype at the locus studied, selective change and population dynamics are independent.

Although selection for melanic forms of moths results from differential predation, the majority of species provide no evidence that abundance has been affected by the change. An exception is the Rosy Minor, *Procus literosa*. This small moth was known in Sheffield many years ago in its typical, rather lichen-coloured, form. Subsequently it disappeared until the area was recolonized by immigrant melanics (Ford, 1964). The sequence of events may be interpreted as arising from increased predation ensuing with change in background colour of the habitat. Predation played a sufficient part in population control for the increased conspicuousness to result in extinction, but when a novel cryptic morph appeared in the area a new ecological balance was established. Selection on the locus is therefore density-dependent. Much of the early discussion of selection on mimetic insects assumes the same to hold true for Batesian mimics, so that reduction in predation leads to a subsequent increase in density. This is by no means necessarily so.

When the selective value of a phenotype does interact with density, it may do so in more than one way. The melanic allele in *Procus* raises the saturation level for the species if increase in its frequency reduces the predation rate. By so doing it allows the population to attain a higher density. There is probably also a strong density-dependent component in searching image behaviour of predators, so that apostatic selection on a polymorphic species permits higher densities than would be possible in a monomorphic species.

Alternatively, a gene which makes the bearer more susceptible to intraspecific competition would increase in frequency when restraints on population density are removed for some reason, and would drop when they again come into effect. It would be difficult to separate these two relations between frequency and density, although one genetic change should just precede the change in density while the other succeeds it.

The higher relative fitness of a genotype can sometimes lower the output in a species. A striking example is the Y-linked male distorter in a species with heterogametic male. Increase in frequency of the distorter gene under selection leads to extinction of the population. The ecological consequence is obviously disadvantageous in this case, but it may not always be so. For example, a balance is probably usually achieved between numbers of host-insect species and the numbers of their parasitoid attackers. The delayed density-dependent relation between the two tends to a steady level, given that the parameters fall within a certain range, and returns to it when the level is disturbed. It was shown many years ago by Nicholson (1933) that under these conditions an increase in the area of search of the parasitoid would decrease the steady level of both species. In the attacking species the trend will undoubtedly be towards increase in the area of search, provided efficiency is unimpaired, because the individual with the greatest searching ability finds the most hosts and, by the definition employed, is the fittest.

By substitution of animals with a large area of search for those with a small one, the mean fitness is raised, while the density is lowered. Similarly, Volterra showed that a mortality factor, such as an indiscriminate insecticide, imposed on an insect predator-prey system could increase the steady level (see Chapman, 1931). If selection occurred for resistance to the insecticide in the predatory species, or in both species, the change in gene frequency could, under these conditions, reduce the density.

Is this a good or a bad thing for the species? In evolutionary terms the fitness of a group of animals is directly related to its biomass. We say that a species is successful if it is very abundant, and shows signs of continuing to be so, or when less abundant, if it displays a high degree of stability. The value of a particular ecology is therefore measured in terms of biomass per unit of time, where the unit is at least some dozens of generations. In the short term the evolutionary tendency in the parasitoid towards greater efficiency of search and lower density is disadvantageous, but it is often true that relatively high densities lead to relatively high levels of instability. This change, where natural selection reduces the density in the short term, could in fact increase the long-term success of the species.

A similar example, where intra-specific selection may have the beneficial effect of limiting numbers has been discussed by Chitty (1960 and elsewhere). The short-tailed field vole, *Microtus agrestis*, like many small mammals of temperate regions, is known to fluctuate in numbers in a cyclic manner. The animals establish territories and fight others which invade their runs, this activity having well-defined and serious physiological consequences, even for the victor. When density is high, individuals in the population are more likely to be in a state of physiological shock or tension than when density is low; and in consequence they suffer reduced output and greater mortality from incidental factors. On its own, however, the effect cannot account for the cyclic fluctuation in numbers. Chitty has suggested that regulation of numbers may be tied up with genetic heterogeneity in the populations. For example, some individuals could be more aggressive than others and more capable of establishing territories, but less robust at high densities as a result of increased levels of physiological stress. Cyclic selection interacting with change in population density could then possibly establish both genetic polymorphism and regular fluctuation in numbers. The observation in the simplified conditions of a *Drosophila* vial population, of a close correlation between density and the frequency of chromosomes bearing the mutant gene *white* has suggested the same interpretation (Thoday, 1963). The importance in population dynamics of the regulatory power of genetic interaction between species, whether herbivore and food-plant, predator and prey or pathogen and host, is discussed by Pimentel (1961).

A class of selection which is sometimes thought to have a deleterious effect on populations is selection of secondary sexual characters. These features, which can be developed to such a remarkable extent, result from competition between males for territory or mates, or from selection by females of a preferred type out of a range of males. It is likely that large structures developed in the context of courtship or territorial competition are a drawback in other phases of life. For example, the branching antlers of some species of deer are (according to Darwin in *The Descent of Man*) less lethal than the short spikes possessed by occasional stags, so that sexual selection may sometimes reduce the survival probability of the animal with a 'good head'. Sexual selection is no different from other types of selection in this respect. Put a stag with large branching antlers in a position of danger from predators and it may be able to flee less easily and fight less effectively than one with smaller antlers. Put the two animals in a position of competition for a harem of females and the better endowed one tends to win. The only place to assess the overall selective value of such characters in wild animals is the total environment in which they normally live. The males with antlers near the modal point in size must then have a net selective advantage over those with larger or smaller ones, and it is very doubtful that the contribution of sexual selection to the balance leads to lower densities or greater instability of numbers than any other kind of selection. In the case of deer it has been suggested (Stonehouse, 1968) that antlers play a part in thermoregulation when in velvet. If so, another component of selection is added which should tend to favour a constant ratio of surface area of antler to body size. If the requirement for thermoregulation in males is related to life in a temperate climate, and this in turn imposes certain patterns of social structure, then the antler is subject to selection as a weapon of defence, as a weapon, real or formal, in courtship, as a temperature regulator and as a social signalling device.

The selection by females of particular male patterns such as the extravagant plumage of pheasants and birds of paradise has always seemed more subtle (Fisher, 1930, compares the opinions of Darwin and Wallace). Fisher suggested that such selection will start with two components, an initial, possibly very small advantage of one kind of male that is due to a natural, not a sexual, advantage, and the additional advantage conferred by female preference. The latter proceeds so long as the sons of the females exerting a choice themselves benefit, and this may go on until well after the trait has reached such extreme development that the initial advantage from natural selection is lost. The progress will continue until checked by some associated disadvantage. Again, the change has no necessary relation to actual numbers or density of the species.

Both types of selection of secondary sexual characters are greatly facilitated if a proportion of each generation fails to mate, so that a chosen mating group determines the genetic composition of the following generation. When almost all adults succeed in mating the association of the chosen pairs has to carry with it a higher than average fitness if the factors responsible are to increase in frequency. In most species a proportion of the adults usually remains unmated, however.

The mathematics of sexual selection has been discussed by O'Donald

(1967, and earlier). The generation of a model requires at least two loci, one controlling the male character selected and the other determining female preference. These loci may be autosomal or sex-linked, although the character is sex-limited. O'Donald finds that sexual selection, unlike natural selection, is much faster for a recessive than for a dominant character, and that selection occurs to favour closeness of linkage between the genes controlling the male character and the female preference. Polymorphism will become established rather readily if natural selection on the character is opposed in direction to the sexual selection.

The relation between relative and absolute fitness values may be summarized as follows. If a locus is picked at random for study there is no reason why it should be found to interact closely with the factors controlling population density, and it is more likely than not that a change in gene frequency under natural selection will be independent of density. If the gene confers a greater capacity for increase on its bearers, then in a stable population they will fill the gaps occurring through mortality but will not affect density. Nevertheless, there will be strong selection in favour of the class of genes that does have effects reacting with density-determining factors, because their contribution to the increased density is equivalent to an increase in relative fitness. The result may be beneficial in terms of survival of the population over a number of generations, but it may be deleterious if the increase in density brings with it an increased instability of numbers. Such instability is very likely to arise if the population relies on a limiting resource, such as the food supply, which shows a lag in recovery when reduced by a large population.

In areas where numbers are determined primarily by factors with density-independent effects, the same argument applies as if there is a density-dependent balance, although decreased stability is much less likely to result from an increase in numbers. Genes raising the rate of increase are likely to be the most effective in raising average numbers in marginal conditions with density independent mortality, whereas genes which increase efficiency of utilization of a resource will be most effective where there is efficient feedback control of numbers. Finally, there is the third class of genes which are favoured over their alleles by selection, but actually decrease the density. These include factors such as the sex ratio distorters and genes acting on systems similar to the host-parasitoid interaction.

Group selection

By raising density, natural selection may decrease stability. A conflict between short-term and longer-term interests therefore arises. It would be good policy for a population to have a high level of genes regulating numbers and controlling exploitation of resources, because the probability of survival would thus be raised. It is therefore attractive to visualize selection taking place at the interpopulational level, where genetically well-buffered populations have a greater survival rate than poorly buffered ones. The selected populations would exhibit limited reproductive output and great sensitivity to density-dependent controlling factors.

The most detailed review of the kind of phenomena involved, which include territory formation (if the territories are larger than the minimum required for survival) and dispersal behaviour, is to be found in the stimulating book by V. C. Wynne-Edwards (1963). Wynne-Edwards writes (p. 20): 'Evolution at this level can be ascribed, therefore, to what is here termed group-selection—still an intraspecific process, and, for everything concerned with population dynamics, much more important than selection at the individual level. The latter is concerned with the physiology and attainments of the individual as such, the former with the viability and survival of the stock or race as a whole. Where the two conflict, as they do when the short-term advantage of the individual undermines the future safety of the race, group-selection is bound to win, because the race will suffer and decline, and be supplanted by another in which antisocial advancement of the individual is more rigidly inhibited.'

Such a state of affairs seems unlikely to persist for long in the face of conventional natural selection. A small group with the ecologically advantageous character would of course be likely to outlive groups without it, but if the character is individually disadvantageous it can only have reached a high frequency by genetic drift. If there was even a small amount of migration between groups, the less well-balanced ones would contaminate their better buffered neighbours, and the genes conferring higher relative individual output at the expense of stability would increase in frequency. Higher animals often show a marked tendency to mate within the group and to exclude immigrants, but in isolation the process would be effected more slowly by mutation. The ecological conditions under which group selection could take place are therefore very restricted. A possible genetical example is the spread of the *t* allele in wild populations of the house mouse (Lewontin and Dunn, 1960).

Another possibility is supplied by the history of the myxomatosis virus after liberation among Australian rabbits (Fenner, 1965). The lethality dropped soon after release from a level of 99·5% kill (average time to death—11 days) to a steady level of 90% kill (average time to death—23 days), apparently because the higher killing power removed the host before the mosquito vector had time to transmit the disease. The new level is one providing optimum conditions for transmission, entailing a balance between virulence, which leads to the presence of infective lesions on the surface of the skin, and comparative benignity, which increases the duration of availability of the host. Selection is obviously taking place between the groups of pathogens infecting different rabbits. If the more virulent viruses have a selective advantage at the individual level within a host, then the overall drop in virulence may be entirely the result of genetic drift and survival or extinction of whole populations of viruses. On the other hand, if the more virulent strain is not fitter than the less virulent one when they are present together on a single host, the mechanism is simply operating at the individual level.

We are left with the problem of interpreting the many examples of ecological and social mechanisms which apparently limit animal species to levels below their saturation densities in terms of other mechanisms. Group selection of population-regulating factors is an attractive general hypothesis. There are some other possibilities, however, that merit close scutiny.

First, populations are limited by the resources available at the least favourable time in the life cycle. It is possible that in some cases of territory formation the social behaviour limits numbers to a level below what can be supported at the time the territory is taken up, but at the maximum that can be tolerated in the severer times to come during the life of the parent and offspring generations involved. The species would then be maximizing its density, but through a mechanism at one remove from the operation of the limiting factor. In this way, the clutch size of birds may be related to the maximum number of young that can be fledged, and not to the larger number of eggs that can be hatched (see Lack, 1966).

Secondly, there is an intermediate class of selection, where the advantage of a gene is immediately expressed, not with reference to the individual acted upon by selection but rather to its close relatives. A well-known example is selection for Müllerian mimicry. A wasp may survive an attack by a predator if it stings its attacker. At the same time it confers a selective value on its colouration if the predator is discouraged from further attacks by the association of colour and disagreeable sensation. This advantage belongs to the individual first attacked and to any of its relatives having a similar pattern. If there is a simple genetic polymorphism for pattern the advantage is therefore high for the siblings of the attacked animal, and falls off with increase in the distance of the relation. The advantage of the gene is partly direct and partly arises in the relatives of the individuals immediately involved. In warningly coloured and distasteful, but relatively delicate, animals such as aposematic caterpillars the larger part of the advantage must accrue to the relatives. The attack is likely to kill the individual involved, however unpleasant the experience for the predator. In fact, aposematic insects are often more robust than their cryptic relatives (Fisher, 1930, quotes Poulton on this subject) so that we might infer either that the direct component is the more important and evolution of warning colouration can only proceed in a species that already has the ability to survive attacks, or that evolution of increased robustness has taken place after the evolution of warning colouration. To distinguish between these possibilities, data on the range of distastefulness are required for species with concealing colouration. Other examples in this category are altruistic behaviour patterns such as the decoying of predators away from their offspring by parent birds, and the development of alarm calls which do not protect the animal alarmed. If the stinging insect in the first example was a bee, which has a comparatively low chance of surviving once it inserts its sting, and which being sterile, does not contribute directly to the forthcoming generation, then it too would be included.

The evolution of this kind of behaviour in social insects has been discussed by Williams and Williams (1957), while Maynard Smith (1964, 1965) dealt with other examples in the category, using the term kin selection to describe them. He distinguishes kin from group selection by pointing out that division of the species into partially isolated units is a favourable but not an essential condition for kin selection, whereas it must occur if group selection is to operate, since the latter depends on the spread of the gene to all members of the group by genetic drift. In order to formalize the conditions favouring kin selection it is necessary to consider the advantage of a gene to the colony as a whole if every individual has the character, and the disadvantage to each bearer. The first of these parameters (A) is the chance of mortality of an individual derived from a group entirely without the gene, divided by the chance in a group composed entirely of the phenotype considered. The second (D) is the probability that a trait-bearer will be killed, in a group comprising 50% of each kind, divided by the chance that a non trait-bearer will be killed. Thus, in the example of distasteful larvae, interest of predators is at first focused on individuals with conspicuous colours. The two categories are A, the mortality from predators in colonies of entirely neutral or cryptic colouration, as a fraction of the mortality of those entirely aposematic in composition, and D, the ratio of mortality of cryptic and aposematic in a colony of 50 : 50 composition.

It is shown both by Williams and Williams and by Maynard Smith in different contexts that the individually harmful trait may be selected provided A is greater than D. The highest genetic resemblance between individuals in a group is found if it is small and consists entirely of siblings. As groups of progressively larger size are considered, so the average number of genes carried in common by two individuals picked at random falls, and the ratio of A to D must increase if the trait is to be selected.

Deductions from hypothetical populations should take account of the fact that real animals live in fluctuating environments. When we say that a trait is individually deleterious but advantageous to the group, the statement involves two time scales. The group advantage arises from the stability of numbers over a series of generations, or from control of output so that resources are used with apparent foresight. The individual disadvantage can be measured within one generation. Over several generations of fluctuating conditions the individual selective value fluctuates, and points may occur where the trait is advantageous. In the simple terms of the fluctuating selective value of a recessive character discussed on pp. 79–81, a character that most of the time is apparently deleterious will be selected provided $\prod w_i > 1$, where i represents a sequence of generations. It is possible that sometimes the environment is sufficiently variable for direct selection of regulating mechanisms to occur, because they materially increase the efficiency of use of resources during a phase of dangerous shortage. The character is in fact favoured by individual selection although it does not appear to be so.

Besides population-regulating factors there are three genetic systems which raise the same difficulties. They are the mechanisms whereby variability is made available to a species, namely mutation, the sexual system and, within the sexual process, the control of recombination. Statements about all of them can have a teleological flavour—they provide sources and stores of variability whereby the population can respond to future environmental change. Natural selection should work, however, to reduce variation to the lowest possible level, because in each generation the variant minority is less fit than the majority. Any hypothesis to account for the genetic control of mutation rate or for the evolution of sex and the genetic control of recombination rate must reconcile the existence of the long-term good with the short-term drawback. The methods available appear to be group selection or individual selection; in this context it is hard to visualize a situation where kin are favoured while an individual carrier is not. Mutations are often recurrent. If a species is established in a constant environment the majority of recurrent mutations will be eliminated as they appear, but the very occasional advantageous one is favoured. A mutator gene such as discussed by Ives (1950) could therefore be selected if the disadvantage of the recurrent mutants was overbalanced by the relative advantage of the occasional favourable mutant. This would only occur in a variable environment where the recurrent mutants sometimes have a chance of proving useful, so that selection could operate to adjust mutation rate to the degree of instability. The rates of selection would be exceedingly small. Kimura (1960) suggests in a difficult paper that the mean mutation rate is adjusted in a much more subtle manner to minimize the genetic load arising from the combination of deleterious mutations and selective substitution.

Whether or not such adjustment occurs there must be a continuous, very slow, creation of beneficial mutants at different times. Fisher and H. J. Muller have pointed out that the importance of the sexual system lies in combining the mutations arising in separate individuals. The rate of evolution in a population composed entirely of asexual organisms is limited by the occurrence of beneficial mutants in clones already embodying a useful mutant. If the mutations appear in two different lineages, both of them increase in frequency at the expense of the non-mutant class, until one is excluded by the other. If, however, there is genetic recombination between individuals the two genes may be incorporated together, and substitution of the double mutant type will take place at a greater speed than that of the fittest single mutant. The advantage of genes controlling recombination therefore depends on the ratio of the time required for a second favourable mutant to occur in a clone already carrying one favourable mutant, and the time required for their combination by the sexual process. Crow and Kimura (1965) have made calculations on this basis, concluding that the greatest advantage of the outbreeding system occurs when populations are large and the environment stable. As Maynard Smith (1968b) points out, this is surprising since the ability to respond rapidly to environmental changes is likely to be critical in small unstable colonies, much less so in large uniform ones.

In extremely unstable conditions the occasional increase in relative fitness of recombinants to a value greater than that of non-recombinants could outweigh the effect of the sequence of generations when selection is against them. It is possible that this is one reason for marginal karyotype monomorphism in flies with many chromosome inversions, the best known examples being *Drosophila willistoni* and *Drosophila robusta*. Carson (1958) selected for positive tactic response to light in *D. robusta*. Two sets of strains were used, derived respectively from a central area of the range where there was a high degree of chromosome polymorphism, and a marginal area where the flies were structurally monomorphic. Selection was imposed for six generations, during which response was more marked in the monomorphic than in the polymorphic lines. Under conditions of selection where a population would benefit from a high level of recombination the chromosome organization which provides it was the more successful. This does not show that the high recombination in the peripheral habitats is a direct result of selection, although under certain circumstances it could be. Rather, we have to account for the structural polymorphism in the central areas. This is related to, and may be a direct response to, the wider variety of habitats utilized, as Dobzhansky has suggested. Whether the range of niches is a cause or an effect, there must be heterosis at some loci and cumulative heterosis between the polymorphic loci within the inversion (Haldane, 1957b). The establishment of chromosomal polymorphism suited to stable conditions therefore has the side effect that when conditions change sufficiently to cause loss of a polymorphism, genetic variability is inevitably made available which will facilitate response to the new conditions. Thus, individual selection leads to development of a system beautifully designed to cope with an unpredictable environment.

9

ESTIMATION OF GENE FREQUENCIES AND
SELECTIVE COEFFICIENTS

It is easier to think of ways in which selection might act than to identify them in practice. The practical procedure usually falls into two parts. It is first necessary to demonstrate a change in phenotype frequency, using a statistical test such as χ^2, and to be satisfied that the observed change supports a hypothesis of selection and is not due entirely to some other process such as change in degree of inbreeding. If the statistical test and the assessment of the observer suggest the action of a particular kind of selective force, the relative size of the force may then be estimated, together with its standard error.

Estimation of selective pressure and of gene frequency are intimately bound up. The frequencies alone may be estimated assuming random assortment and equal viability. It is usually difficult, however, to decide when these assumptions are justified. In addition, it is often impossible to infer anything about selection from phenotype ratios when some genotypes are indistinguishable. Estimation of the gene frequency on the basis of equal viability when there are several alleles showing dominance is something of an act of faith.

The method of estimation

In Chapter 1 we introduced the problem of the probability of getting a head or a tail when a coin is spun. Defining these probabilities as p and $q(=1-p)$ the chance of n heads in a row is p^n, and the chance of a sequence of n tails is q^n. A mixture of heads and tails in the sequence has a probability of occurrence of $1 - p^n - q^n$. If p and q equal $\frac{1}{2}$ and the coin is thrown twice the chance of two heads is $\frac{1}{4}$, of one head and one tail is $\frac{1}{2}$, and the chance of two tails is $\frac{1}{4}$. The total must always be unity since one or other side of the coin is revealed on each throw. The probability of the mixed sequence in two throws is the sum of the probabilities of the sequences (head, tail) and (tail, head), each of which is $pq = \frac{1}{4}$. When there are ten throws the probability of five heads is p^5q^5 raised by a factor that takes account of all the permutations in which five heads and five tails can appear. It may readily be shown that the probability of any number of heads a, out of a sequence of n throws is

$$\frac{n!}{a!(n-a)!}\, p^a q^{n-a}$$

The right-hand part, consisting of a product of two probabilities raised to the powers of the number of times of occurrence, is the probability of any specific sequence of a occurrences in a total of n. The left-hand part containing the factorials is the number of such sequences. When a is equal to n this expression reduces to p^n, as we should anticipate. When several possible outcomes are open with probabilities p, q, r, s, etc., summing to 1, the function giving the probability of a, b, c, d, etc., events in a total of n is

$$\frac{n!}{a!\,b!\,c!\,d!\,\cdots}\, p^a q^b r^c s^d \cdots$$

Now, if in fact we try the experiment of throwing a coin repeatedly, and obtain 8 heads in 12 throws, we can draw three kinds of conclusion about p and q. The first is that the probability of obtaining a head is truly one half, the probability of getting 8 heads in 12 throws, which is 0·12, being insufficiently small to force us to reject this belief. On the other hand, we may reject entirely the prior assumption that the probability is one-half if we feel that it is invalidated by the evidence of the frequency with which the coin falls face upward. This would be an appropriate course if no information was available about the possible direction and size of the bias, if any, possessed by the coin. In that case the best available estimate of the probability of getting a head is 8/12, the frequency with which heads occurred in the run. Which choice is preferred is likely to be influenced by the size of the experiment. Twelve is a small number of throws. We should be a lot more likely to adopt the second course if 800 heads occurred in 1200 than when 8 occur in a total of 12.

The third course would be to use the results of the trial run to modify the prior belief in some way. We would then start with an idea of what should happen $(a/n = \frac{1}{2})$ and balance against it evidence of what has happened in the trial $(a/n = \frac{3}{4})$, to give an estimate between these two values at a point determined by the relative weight given to the two kinds of evidence. The third course of action appears to be the most appropriate, but the one where all prior beliefs are discarded in favour of an estimate based entirely on the experimental data is the one almost always used in estimation problems in genetics. This procedure is employed in the method of maximum likelihood. When applied to an elementary example the choice may seem obtuse, and it is worth looking at a slightly more complex estimation from the related field of population ecology in order to appreciate the reasons for it.

When the size of a mobile population is estimated by mark-release-recapture, three statistics are obtained: the size of the sample marked and released, the size of the second sample and the number of marked individuals in the second sample. Calling these values r, s, and m, and the unknown total population size N, we have, on the basis of random selection of individuals from the population, $\frac{r}{N} = \frac{m}{s}$, so that $N = \frac{rs}{m}$. This is the maximum likelihood estimate of the population size (Bailey, 1951). Having fixed r in the first stage of the procedure the possible outcomes of the second sampling may vary between the limits $m = 0$ and $m = s$. The estimated population size may therefore vary from $N = r$ to $N = \infty$. All values over this range are regarded as equally acceptable, just as in the first example any value of p between 0 and 1 would have been acceptable. But the mark-recapture method is normally used only when the experimenter thinks that $N = r$ is most unlikely; and knowledge of the ecology of populations usually causes his willingness to accept the estimate to diminish steadily as it tends towards infinity. A value of the order of about $3r$ to $30r$ is usually envisaged when recapture methods are employed, the subjective acceptance falling steeply below this modal region and more gradually above it. Now, if we are able to make some sort of prior assessment of N, such as the one in the preceding sentence, there would be some virtue in incorporating it in the final determination. Parr, Gaskell and George (1968) discuss the method for doing so for population estimations, although they do not recommend any particular equation. It appears, however, that the reasoning involved when incorporating a prior probability function does not necessarily lead to the same estimate of N as is provided by the maximum likelihood equation.

As the number of factors determining the estimated parameter becomes more numerous and complex, so the difficulty of knowing whether the prior assessment is justified increases. We can be more certain that a coin is unbiased than that N is a lot larger than r. Inspection of the shape and weight of a coin tells us a good deal about how it will behave when spun, but the number of insects in a hedgerow is extremely difficult to assess. One reason why the

maximum likelihood method is favoured by most statisticians working
in biology is that as a general method the limitation of the *a posteriori*
approach is outweighed by the avoidance of serious error that may
be introduced by making the wrong prior probability assumptions.
It is interesting to read the remarks of Yates, Lindley and Fisher with
reference to the choice of method of approach to genetical problems,
for example, whether the estimation of linkage distance between two
loci should be influenced by knowledge of the chromosome number.
(Yates, 1958, comments on a review by Lindley of Fisher's book,
Statistical Methods and Scientific Inference.) The maximum likelihood method, which was first developed by Fisher, has other important properties. It is quite easy to apply—much more so than methods
involving prior probability assumptions—and the variance of the
estimate is always obtainable. When several similar methods of
estimation can be thought of, which use the basic data in slightly
different ways, the method of maximum likelihood provides the
smallest variance for the estimate. That is, it extracts from the data
the greatest amount of information about the estimate, and in this
statistical sense, is said to be fully efficient. The mathematics of the
method is well described by Mather (1965) and, at a more advanced
level, by Bailey (1961).

The maximum likelihood method

In order to find the maximum likelihood estimate of p, given 8 heads
in 12 throws, we proceed as follows. The probability of this event is

$$\frac{12!}{8!\,4!}\,p^8(1-p)^4$$

where p is the probability of getting a head. This function, called the
likelihood function, has a very low value when p is close to zero or
close to 1, and reaches a peak somewhere between. At the maximum
the number of heads to be expected is closest to the number observed
(in this case the two are equal although this is not always so). The
value of p for which the likelihood is a maximum is found by differentiating the function with respect to p and equating to zero. In
doing so the factorial part is ignored, since its value does not affect
the position of the mode, and the remaining part is converted to
natural logarithms in order to simplify differentiation. This, too,
leaves the mode unaffected. The log likelihood function is therefore,

$$L = 8 \log p + 4 \log(1 - p)$$

so that (since $d \log x = dx/x$),

$$\frac{dL}{dp} = \frac{8}{p} - \frac{4}{1-p}$$

$$= \frac{8 - 12p}{p(1-p)}$$

At the required maximum $8 - 12p = 0$, so that $p = \frac{8}{12}$.

The variance of maximum likelihood estimates is found from the relation

$$V_p = - \left(\frac{d^2L}{dp^2}\right)^{-1}$$

the expected number being substituted for the number observed, after differentiation. This is an accurate estimate of the variance provided the sample on which it is based is reasonably large—as a rough guide, in practice, say for totals (n) of more than fifty. Proceeding with the present small-scale example we have the expected number $np = a$, and

$$\frac{dL}{dp} = \frac{a - np}{p(1-p)}$$

The derivative is

$$\frac{-a + 2ap - np^2}{p^2(1-p)^2}$$

or, after substitution of np for a, $- \frac{n}{p(1-p)}$. The variance is accordingly $\frac{p(1-p)}{n}$, so that the standard error of p is $\sqrt{\left(\frac{p(1-p)}{n}\right)}$, an expression which will be familiar as the binomial standard error. The standard error cannot always be used directly to attach confidence intervals to p, because as the estimate of p moves away from 0·5 in either direction its probability distribution becomes increasingly asymmetrical, tending at the extremes to a Poisson distribution. Some kind of conversion is therefore required to find the upper and lower limits for a given probability. In the simple binomial case, for example, this may be done by the use of Table VIII in Fisher and Yates (1963).

In more complex cases it may be impossible to solve the log likelihood equation explicitly. Even when it is possible, the calculation of the variance involves differentiation of a quotient, which may lead to cumbersome algebra. Both difficulties may, however, be circumvented. A value of the unknown parameter p may be found to any desired accuracy by calculating dL/dp with trial values of p substituted in the function, and plotting the resulting values of dL/dp on p. In this way a curve is produced which cuts the p axis at $dL/dp = 0$, the required estimate of p. Then, since d^2L/dp^2 is the gradient of the curve, the variance may be estimated by finding the change in dL/dp about $dL/dp = 0$ over as short a change in p as is practical.

The relation of the curve of dL/dp to p is also the basis of a method of successive approximation for finding the value of p. This may be used when the algebraic expressions dL/dp and d^2L/dp^2 have been found, but cannot be solved for p. Instead, an arbitrary value p_0 is put into the expressions. The calculated value of dL/dp is then the height of the curve above $dL/dp = 0$ when $p = p_0$ and d^2L/dp^2 is the gradient of the curve as before. Consequently, a new value p_1 can be found which is a better estimate of p, because

$$(dL/dp)/(p_0 - p_1) \simeq d^2L/dp_0^2$$

Rearranging this equation we find

$$p_1 = p_0 + \left(\frac{dL}{dp_0}\right) V_{p0}$$

The process may then be repeated to find p_2, p_3, etc., until two successive values very close to each other are produced. These are then good estimates of the maximum likelihood value. Provided the appropriate probability function can be derived, estimates of mean and variance are therefore obtainable by one method or the other, without fear of defeat by the mathematics.

Estimation of gene frequency

The estimate derived above, and its variance, applies equally whether we are discussing the frequency of heads and tails or the gametic frequencies of a pair of alleles. The numbers in the two classes of gamete are the basic data, instead of the numbers of heads and tails. The values of p and q are then the estimated gene frequencies. Moving on to populations of diploids the same procedure is followed. When there is dominance a sample from the population contains a individuals of the dominant phenotype in a total of n. The two phenotype frequencies are $(1 - q^2)$ dominants and q^2 recessives. The log likelihood function is therefore

$$L = a \log(1 - q^2) + (n - a)\log q^2$$

so that

$$\frac{dL}{dq} = -\frac{2aq}{1-q^2} + \frac{2q(n-a)}{q^2}$$

$$= \frac{-2aq + 2(1-q^2)(n-a)}{q(1-q^2)}$$

The estimate obtained by equating this expression to zero is $q = \sqrt{\left(\frac{n-a}{n}\right)}$. The variance may be found in the same way as before, or by means of a slightly different calculation which is sometimes easier to use. The general form of all the log likelihood expressions used in these estimations to find a parameter q is $L = \sum a_i \log m_i$ where the m_i represent the frequencies and the a_i represent the numbers in each class. The second derivative d^2L/dq^2 is equal to $-n \sum \frac{1}{m_i}\left(\frac{dm_i}{dq}\right)^2$, where $n = \sum a_i$. In the present example we therefore have

$$\frac{1}{V} = n\left[\frac{1}{(1-q^2)}(-2q)^2 + \frac{1}{q^2}(2q)^2\right] = \frac{4n}{1-q^2}$$

The standard error of q is accordingly $\sqrt{\left(\frac{1-q^2}{4n}\right)}$.

A sample of the moth *Gonodontis bidentata* taken in Manchester in 1968 consisted of 79 individuals of the dominant melanic morph and 21 typicals. The value of q^2 is therefore 0·21, so that the gene frequency of the typical is $\sqrt{(0\cdot21)} = 0\cdot458$, with a standard error of $\sqrt{(1 - 0\cdot21)/400} = 0\cdot044$.

In Table 9.1 these expressions are listed, together with others for the case of two autosomal alleles with no dominance, for sex linkage, for three autosomal alleles in a dominance series, and the general dominance series. As a numerical example of a 3-allele case we may consider data collected in Manchester for the other striking polymorphic moth, *Biston betularia*. There are three morphs, the black *carbonaria*, the darkly mottled *insularia* and the pale typical. Their appearance is controlled by three alleles, the *carbonaria* allele being dominant to both the others, and *insularia* being dominant to typical. There may in fact be several alleles with intermediate effects (Clarke and Sheppard, 1964; Lees, 1968) but all the *insularia* in the sample had a very similar appearance and were probably controlled by the same gene. The total sample contained 521 *carbonaria*, 4 *insularia* and 13 typicals. Using the expressions in Table 9.1 we get 0·822 ± 0·021,

0·022 ± 0·030 and 0·155 ± 0·021 for the three gene frequencies. With these results it would be incorrect to treat the standard error as if it referred to a symmetrical distribution. For the typicals, for example, the upper 95% confidence interval is considerably more than two standard errors above the mean, and the lower limit is less than two standard errors below it. A serious intrinsic source of error arises from the fact that the estimation depends for its accuracy on the correctness of the assumptions about relative frequencies of homozygotes and heterozygotes within the phenotype classes. The *carbonaria* phenotype is composed of three genotype classes and the *insularia* phenotype of two classes. The sources of deviation from expectation are the effect of the breeding structure of the population and the effect of selection. If there is a high degree of inbreeding the level of homozygosity is raised. (Yasuda and Kimura, 1968, apply a factor to take account of inbreeding or selection bias in the estimation of gene frequencies, but the properties of a dominance series of this kind make it impossible to estimate the factor.)

In *betularia* there is unlikely to be much distortion of genotype ratios as a result of inbreeding since it is a fairly mobile moth. This is not so, however, for the previous example, *Gonodontis bidentata*, which appears to be more sedentary in behaviour. In both species, the genotype frequencies are likely to be disturbed by selection—indeed they must be so to some extent if the alleles coexist at stable equilibria under frequency-independent selection. Direct evidence indicating selection was obtained from gene frequency estimations by Murray (1966), for what is effectively a four-allele system in *Cepaea nemoralis*. The genes controlling shell ground colour and banding are very closely linked, so that when pink and yellow ground colours are present there are four kinds of chromosome—pink unbanded, pink banded, yellow unbanded and yellow banded. Pink is dominant to yellow and unbanded dominant to banded, so that calling the four chromosome frequencies p, q, r and s we have the following situation.

type	pink unbanded	pink banded	yellow unbanded	yellow banded
observed number	b	c	d	e
expected number	$(p^2 + 2pq + 2pr + 2ps + 2qr)n$	$(q^2 + 2qs)n$	$(r^2 + 2rs)n$	s^2n

Proceeding as in the three-allele *betularia* example, the frequencies are estimated as

$$p = 1 - (q + r + s), \quad q = \sqrt{\left(\frac{c + e}{n}\right)} - \sqrt{\left(\frac{e}{n}\right)}$$

$$r = \sqrt{\left(\frac{d + e}{n}\right)} - \sqrt{\left(\frac{e}{n}\right)} \quad \text{and} \quad s = \sqrt{\left(\frac{e}{n}\right)}$$

Murray calculated the chromosome frequencies for a series of random samples from the Isles of Scilly, some of which provided negative frequencies. For example, the results for two samples are

frequency	p	q	r	s
Tresco, Main	−0·017	0.749	0.078	0·190
St. Warna, West	−0·036	0.584	0.083	0·369

This shows clearly that the assumptions about the genotype composition of the phenotype classes are incorrect, most probably as a result of selection.

When series of more than three alleles are considered the maximum likelihood method becomes difficult to handle because of the variety of possible scorable phenotypes which may be encountered. With three alleles there are six possible diploid genotypes, with four there are ten and with n alleles there are $\frac{1}{2}n(n + 1)$ genotypes, many of which are usually masked by recessiveness of the effects of some alleles to others. Each phenotype combination for a given number of alleles has to be analyzed in a different way. Cotterman (1953) has studied and classified the systems, using the term phenogram for the formula describing a particular system. The phenogram consists of a sequence of three numbers: for example the sixth entry in Table 9.1 has the phenogram 3–3–6 meaning that it is a system of three alleles and three phenotypes of the sixth kind. The last number is an arbitrary one, required because there are 21 possible three-allele, three-phenotype systems. There are in all 52 different possible three-allele combinations. With the addition of another allele both the number of permutations and the amount of computation is greatly increased. An iterative method has been developed which overcomes the difficulties of direct estimation when there are several alleles (Ceppellini, Siniscalco and Smith, 1955; Yasuda and Kimura, 1968). The method starts from the expected genotypic composition of phenotypes, based on the assumption of Hardy–Weinberg ratios, but allows modification to take account of discrepancies which may appear in some phenotypes. As with the procedure already outlined, it is suitable for the determination of gene frequencies in cases such as human blood-group data derived from large random mating populations, when

Coefficients of Natural Selection

there is little evidence of selective distortion of genotype frequencies.

The method—called gene counting—may be exemplified by returning to the question of the frequency of typicals in a sample of the moth *Gonodontis bidentata*. There are 79 individuals of the dominant melanic phenotype $(M-)$ and 21 typical (mm) individuals. As a first estimate of gene frequency, use an arbitrary value for q, the frequency of m, such as $q_0 = 0.5$. Then on the basis of Hardy–Weinberg ratios, the melanic phenotype consists of $p^2/(p^2 + 2pq)$, or one-third, of MM individuals and $2pq/(p^2 + 2pq)$, or two-thirds, of Mm individuals. The three genotypes are therefore represented in the sample by 26, 53 and 21 individuals respectively. On this basis the frequency q_1 is $\dfrac{53 + 2 \times 21}{200}$, or 0.475. Repeating the process with the new estimate of gene frequency we find $q_2 = 0.465$. Successive values converge on the final best estimate of q, and the process of substitution and repetition may be stopped when two successive values are sufficiently close to meet some desired level of accuracy. In this case the value to which the estimates converge is the one obtained from the direct calculation, namely 0.458. When there are more alleles and phenotypes this is not always so, however. For example, when Murray's data for *Cepaea nemoralis* are recalculated using the gene counting method they provide the following estimated frequencies, which may be compared with the ones on p. 181. These series are better estimates

frequency	p	q	r	s
Tresco, Main	0.0000	0.7383	0.0613	0.2003
St Warna, West	0.0000	0.5576	0.0474	0.3950

than the first of the true gene frequencies, and looking at the results there is now no indication that the assumptions of Hardy–Weinberg equilibrium are not exactly met. The main advantage of the gene-counting method is that it is equally applicable to more complex cases and especially suitable for use with the computer.

Table 9.1

Estimated frequencies and their variances for different genetic systems in the absence of distortion by selection and inbreeding. In part after Cotterman (1954) and Yasuda and Kimura (1968).

(a) *2 gametes, or morphs or hemizygotes*

type	L	M
observed number	b	c
expected number	pn	qn

Table 9.1—*continued*

$$p = \frac{b}{n} \qquad q = \frac{c}{n} \qquad \text{Var} = \frac{pq}{n}$$

(b) *2 alleles, all genotypes distinguishable*

type	AA	Aa	aa
observed number	b	c	d
expected number	$p^2 n$	$2pqn$	$q^2 n$

$$p = \frac{2b + c}{2n} \qquad q = \frac{2d + c}{2n} \qquad \text{Var} = \frac{pq}{2n}$$

(c) *2 alleles, dominance*

type	$A-$	aa
observed number	b	c
expected number	$(1 - q^2)n$	$q^2 n$

$$p = 1 - q \qquad q = \sqrt{\left(\frac{c}{n}\right)} \qquad \text{Var} = \frac{1 - q^2}{4n}$$

(d) *2 alleles, sex linkage, all genotypes distinguishable*

type	AA	Aa	aa	AY	aY
observed number	b	c	d	e	f
expected number	$p^2 n$	$2pqn$	$q^2 n$	pn'	qn'

$$p = \frac{2b + c + e}{2n + n'} \qquad q = \frac{2d + c + f}{2n + n'} \qquad \text{Var} = \frac{pq}{2n + n'}$$

(e) *2 alleles, sex linkage, dominance*

type	$A-$	aa	AY	aY
observed number	c	d	e	f
expected number	$(1 - q^2)n$	$q^2 n$	pn'	qn'

$$p = 1 - q \qquad q = \frac{-e + \sqrt{[e^2 + 4(2d + f)(2n + n')]}}{4n + 2n'}$$

$$\text{Var} = \frac{q(1 - q^2)}{4nq + n'(1 + q)}$$

(f) *3 autosomal alleles, dominance series*

type	$A'-$	$A-$	aa
observed number	b	c	d
expected number	$(p^2 + 2pq + 2pr)n$	$(q^2 + 2qr)n$	$r^2 n$

$$p = 1 - \sqrt{\left(\frac{c + d}{n}\right)} \qquad \text{Var}_p = \frac{p(2 - p)}{4n}$$

$$q = \sqrt{\left(\frac{c + d}{n}\right)} - \sqrt{\left(\frac{d}{n}\right)} \qquad \text{Var}_q = \frac{q[2 - q(1 - p)]}{4n(1 - p)}$$

Table 9.1—*continued*

$$r = \sqrt{\left(\frac{d}{n}\right)} \qquad\qquad \text{Var} = \frac{1 - r^2}{4n}$$

(g) *m autosomal alleles, dominance series*

type	$A_1A_1 \cdots A_1A_m$	$A_2A_2 \cdots A_2A_m \cdots A_mA_m$
observed number	n_1	n_2 $\cdots$ n_m
expected number	p_1n_t	p_2n_t $\cdots$ p_mn_t

$$\sum n_i = n_t \qquad N_i = n_i + \cdots + n_m \qquad P_i = p_i + \cdots + p_m$$

$$p_i = \sqrt{\left(\frac{N_i}{n_t}\right)} - \sqrt{\left(\frac{N_{i+1}}{n_t}\right)}$$

$$\text{Var } p_i = \frac{p_i(2 - p_i) + 2p_i(1 - P_i)/P_i}{4n_t}$$

Estimation of selective values

The Δq equations used in previous chapters describe the change in gene frequency occurring under particular conditions of starting frequency and selection. In principle, it is possible to start with information on q and Δq and calculate the selective values involved. For example, if the fitnesses of three genotypes are $1 : 1 : 1 - s$, so that $\Delta q = -sq^2(1 - q)/(1 - sq^2)$, then rearrangement provides $s = -\Delta q/q^2(1 - q - \Delta q)$, or, since $\Delta q = q_1 - q_0$, $s = (q_0 - q_1)/q_0^2 (1 - q_1)$. If we had a situation where two successive gene frequencies were 0·4 and 0·5, then $q = 0.4$ and $\Delta q = 0.1$, so that s would be -0.125. This procedure is appropriate provided the fitness of the genotype at frequency q^2 really is recessive, as the equation implies, and that the three genotypes are distinguishable so that the two values of q in the samples may be found. It is also necessary for the samples to be reasonably large, so that the errors attached to the values of q are small.

When two coefficients s_1 and s_3 have to be estimated relative to the third, a Δq equation for a single initial value of q will provide no more than the ratio of s_1 to s_3. This is seen most directly when $\Delta q = 0$, at which point (Table 1.4), $s_1 = s_3q/p$. To provide estimates of s_1 and s_3 relative to the third coefficient it is necessary to have two Δq equations for different values of q, which may then be solved as a pair of simultaneous equations. The natural extension of this procedure is to take a series of data for several generations and fit a line to the regression of Δq on q. If there is evidence of heterozygote advantage, an estimate of the slope of Δq on q may be obtained by determining the linear regression line in the region of the equilibrium. The point where the line cuts the $\Delta q = 0$ coordinate indicates the ratio of the selective coefficients of the homozygotes, and the calculated regression coefficient may be equated to the expression for $d\,\Delta q/dq$ at the equilibrium value of q to give an estimate of their magnitudes. Although this method is very rough, it will indicate the approximate size of the pressures involved. In fact, of course, the relation of Δq to q is curved, and better results will be obtained by the iterative fitting of a selection curve. A procedure, using the method of least squares, is described by Wright and Dobzhansky (1946). In the notation of these authors the observed values of Δq are signified by y, while $x_1 = pq^2/\overline{w}$ and $x_2 = p^2q/\overline{w}$. We shall call the selective coefficients of the two homozygotes s_1 and s_3, that of the heterozygote being zero. If first approximations to s_1 and s_3 are used to calculate $\overline{w}$ for each observed value of q and Δq, new estimates of the coefficients are obtained from the equations.

$$(\textstyle\sum x_2^2)s_1 - (\sum x_1 x_2)s_3 = \sum x_2\,y$$
$$-(\textstyle\sum x_1 x_2)s_1 - (\sum x_1^2)s_3 = -\sum x_1 y$$

Values of $\overline{w}$ are recalculated using the means of the old and the new estimates of s_1 and s_3, and the process of substitution and recalculation is repeated until successive values of the coefficients are sufficiently close. If one allele is visually dominant to the other but there is evidence that the fitnesses of the three genotypes differ, Δq and q will not be known. The best procedure then is probably to estimate q as the square root of the recessive morph frequency, carry out the calculation and repeat it using new estimates of q derived from the combined evidence of morph frequency and first estimates of s_1 and s_3.

Another situation which may arise is one where frequencies are known at two points in time separated by several generations. The procedure for estimating the selective coefficient responsible for the change by integration of a Δq equation is given on pp. 48–50. As pointed out there, Clarke and Murray (1962) provided reasonable limits for the selection acting on different morphs of *Cepaea nemoralis* in this way, but the method is a mathematically approximate one, and it depends on several assumptions such as the constancy of the selective pressures over the intervening period, about which little evidence may be available.

Discussion of estimation procedures has been concerned with constant selection pressures, whereas it is emphasized in previous chapters that selection may act in several other ways. Wright and Dobzhansky show that their experimental results may be fitted almost equally well by curves based on the assumption of simple constant selection, of constant selective pressures differing between the sexes and of frequency-dependent selection. Sometimes, however, distinction between the mode of selection can be made from data on change in gene or morph frequency. For example, when they studied alleles at the Esterase–6 locus in *D. melanogaster*, Kojima and Yarborough (1967) showed that the estimated relative viabilities varied markedly with change in the gene frequency at which parental populations were set up. They therefore concluded that the selective agent acted in a frequency-dependent manner. In another instance, Williamson (1960) pointed out that for the *medionigra* gene in the Cothill population of *Panaxia dominula* the slope of Δq on q about the equilibrium is too steep to be the result of heterosis with constant selection. The polymorphism must therefore be maintained by some other kinds of effects. As a rule, however, the nature of the selection must be inferred from independent evidence, and only the size of the force on the most probable hypothesis of action, from the observed change in gene frequency.

Apart from assumptions about how selection is acting the methods described involve several problems, such as the estimation of gene frequencies when there is dominance and estimation of the size of the sampling errors involved. Decisions have to be made on how they should be treated, and the resulting estimate may vary with the choice of treatment. For this reason a common approach to all problems of estimating selection is desirable. The use of maximum likelihood would be appropriate, although so far the field has received comparatively little attention from statisticians, compared with gene frequency estimation. In what follows, an outline of the problem in maximum likelihood terms is given. It should be emphasized that it is no more than an outline, designed to illustrate the general methods applied, and that more detailed treatment may be required for any specific experimental example.

The most straightforward cases are ones where the initial gene or morph frequency is known in an experimental investigation. A simple problem of this kind, mentioned in Chapter 1, concerns the estimation of net selective pressure from a change in frequency of two morphs. Non-mimetic and mimetic insects are released at frequencies L and M. They are subject to selective predation, so that a sample of n recaptures contains b mimics and a non-mimics. If w is the selective value of the non-mimic the likelihood function is

$$\left(\frac{wL}{wL + M}\right)^a \left(\frac{M}{wL + M}\right)^b$$

Differentiating the logarithm of this expression we get

$$\frac{dL'}{dw} = \frac{aM - bwL}{w(wL + M)}$$

so that

$$w = \frac{aM}{bL}$$

Further calculation shows the variance to be

$$V_w = \frac{w\overline{w}^2}{LMn}$$

where $\overline{w} = wL + M$. In one part of the mimicry experiment referred to on p. 37 a total of 634 insects were released at frequencies of 0·503 non-mimics and 0·497 mimics. One hundred and sixty-five insects were recaptured, of which 72 were non-mimics. The difference in frequency between release and recapture is just formally significant using the χ^2 test ($P = 0·05$). Accepting this level of probability as indicating selective elimination we may proceed to estimate the size of the selective pressure. Substitution in the equations above provides an estimate of w for the non-mimics of 0·765, with a standard error of 0·120. The selective disadvantage is 23·5%. This result is in good agreement with the test applied to detect selection, because the estimated value is just two standard errors below the value of 1 that would indicate equality between the morphs. Woolf (1955) and Edwards (1965) discuss similar cases of selective elimination, using the logarithm of w, which has a symmetrical distribution about unity. Woolf described a significance test for use when several estimates are obtained from independent observations of varying sample size.

When we move on to populations of diploids there are two selective values to be estimated relative to the third. To give an example of the case where all genotypes are distinguishable, an artificial colony of the scarlet tiger moth *Panaxia dominula* was established by releasing eggs from crosses of parents heterozygous for the *medionigra* gene (Sheppard and Cook, 1962).

On the assumption of Mendelian segregation with no selection the subsequent gene frequency should be unchanged from the 50% released, and there should be a 1 : 2 : 1 genotype ratio. In fact, the sample collected contained 36 typicals, 42 *medionigra* heterozygotes and 21 individuals of the homozygous form *bimacula*, a distribution which differs significantly from the expected ratio. When there is selection the expected frequencies in each class are

$$\frac{w_1 p^2}{\overline{w}}, \quad \frac{2pq}{\overline{w}}, \quad \frac{w_3 q^2}{\overline{w}}$$

where q is the frequency of the *medionigra* gene and $\overline{w} = w_1 p^2 + 2pq$

$+ w_3 q^2$. The probability of a specific sequence giving the observed numbers is therefore

$$\left(\frac{w_1 p^2}{\overline{w}}\right)^a \left(\frac{2pq}{\overline{w}}\right)^b \left(\frac{w_3 q^2}{\overline{w}}\right)^c$$

In order to estimate w_1 and w_3 the logarithm of the likelihood expression must be differentiated, first with respect to w_1 and secondly to w_3. The resulting derivatives are then equated to zero to give a pair of simultaneous equations that may be solved for w_1 and w_3. For the first,

$$\frac{\partial L}{\partial w_1} = \frac{a\overline{w} - nw_1 p^2}{w_1 \overline{w}} = 0$$

and similarly,

$$\frac{\partial L}{\partial w_3} = \frac{c\overline{w} - nw_3 q^2}{w_3 \overline{w}} = 0$$

Solution of the two equations provides

$$w_1 = \frac{2aq}{bp}$$

and

$$w_3 = \frac{2cp}{bq}$$

When we put $p = q = \frac{1}{2}$, and substitute the numbers of the three genotypes of the moth for a, b and c, the values obtained are $w_1 = 1\cdot714$ and $w_3 = 1\cdot000$. The influence of the gene on survival is effectively dominant so far as these data are concerned, the three fitnesses being $1\cdot71 : 1 : 1$, or $1 : 0\cdot58 : 0\cdot58$.

The variance of w_1 is now influenced to some extent by the accuracy with which we know w_3. Both estimates are therefore involved in the determination of the variance of each. The method of finding the variances, which involves inverting the matrix of second derivatives of the likelihood functions, is described by Bailey (1961). The second derivatives are $I_{11} = -\partial^2 L/\partial w_1^2$, $I_{33} = -\partial^2 L/\partial w_3^2$ and $I_{13} = I_{31} = -\partial^2 L/\partial w_1 \partial w_3$. Using them, we estimate the variances as

$$V_{11} = -I_{33}/(I_{13}^2 - I_{11} I_{33})$$

$$V_{33} = -I_{11}/(I_{13}^2 - I_{11} I_{33})$$

and the covariance of w_1 and w_3 as

$$V_{13} = I_{13}/(I_{13}^2 - I_{11}I_{33})$$

In the case discussed, where all genotypes are distinguishable, simplification of the resulting expressions provides

$$V_{11} = \frac{\overline{w}(w_1 p + 2q)w_1}{2p^2 qn}$$

$$V_{33} = \frac{\overline{w}(w_3 q + 2p)w_3}{2pq^2 n}$$

and

$$V_{13} = \frac{\overline{w} w_1 w_3}{2pqn}$$

For the numerical example, the standard error of w_1 obtained by this means is 0·389, and the standard error of w_3 is 0·267.

When several independent experiments have been carried out, perhaps with different initial gene frequencies, the joint estimation of fitnesses may be made by summing the log likelihood expressions for each experiment and equating the sum to zero. These equations cannot be solved directly, and the best procedure when two parameters are estimated is the method of successive approximation. Values guessed to be close to the true values are chosen for w_1 and w_3. Second estimates w_1' and w_3' are then obtained from the equations

$$w_1' = w_1 + V_{11}\left(\frac{\partial L}{\partial w_1}\right) + V_{13}\left(\frac{\partial L}{\partial w_3}\right)$$

$$w_3' = w_3 + V_{31}\left(\frac{\partial L}{\partial w_1}\right) + V_{33}\left(\frac{\partial L}{\partial w_3}\right)$$

When several sets of data are combined, both the first derivatives and the I's, from which variances and covariances are obtained, are the sums of the individual values for each set. The new values of w_1 and w_3 are used to recalculate $\partial L/\partial w_1$ and $\partial L/\partial w_3$, and the process is repeated until the difference between successive values is negligible. The variances and covariances change slowly so that it is not essential to recalculate them at each round.

In the example of the scarlet tiger moth the size of two fitnesses relative to the third is estimated from the numbers of the three genotypes present in a sample. Two of these quantities may vary independently, the third being fixed by the other two. The gene frequency at the outset was known. This is essential to the method because the number of parameters that may be estimated cannot exceed the number of independently varying categories in the sample. Had the starting gene frequency been unknown, it might at first appear possible to find it, too, by differentiation of the log likelihood function with respect to q. We should then have had three simultaneous equations but only two observed independent relations—a/n and b/n. The equations could not be solved for w_1, w_3 and q together. When the scored effect of one allele is dominant to that of the other there are but two phenotype classes and one independent variable. The selective values cannot be calculated in the way outlined unless there is assumed to be dominance in fitness as well as in visual expression, an assumption that is not usually justified.

The limitation may be overcome by a method used by Creed (1963) involving data from two generations. Suppose that a_0 dominants and b_0 recessives are observed in one generation ($a_0 + b_0 = n_0$), and a_1 and b_1 in the subsequent generation ($a_1 + b_1 = n_1$). Two likelihood functions are involved, viz.:

$$\left(\frac{w_1 p_0^2 + 2p_0 q_0}{w_1 p_0^2 + 2p_0 q_0 + w_3 q_0^2}\right)^{a_0} \left(\frac{w_3 q_0^2}{w_1 p_0^2 + 2p_0 q_0 + w_3 q_0^2}\right)^{b_0}$$

for generation 0, and

$$\left(\frac{w_1 p_1^2 + 2p_1 q_1}{w_1 p_1^2 + 2p_1 q_1 + w_3 q_1^2}\right)^{a_1} \left(\frac{w_3 q_1^2}{w_1 p_1^2 + 2p_1 q_1 + w_3 q_1^2}\right)^{b_1}$$

for generation 1. The information contained in them may be combined by writing the second expression entirely in terms of the known ratio $a_0 : b_0$ and of the parameters to be estimated. Let $B = n_0/b_0$. Then it follows that

$$B = \frac{w_1 p_0^2 + 2p_0 q_0 + w_3 q_0^2}{w_3 q_0^2}$$

Rearrangement of this equation provides,

$$q_0 = \frac{w_1 - 1 - \sqrt{1 - w_1 w_3 (1 - B)}}{w_1 + w_3 - 2 - Bw_3}$$

Now,

$$q_1 = \frac{w_3 q_0^2 + p_0 q_0}{w_1 p_0^2 + 2p_0 q_0 + w_3 q_0^2}$$

and $p_1 = 1 - q_1$, so that although the result is inconvenient to handle, the second function may be rearranged to contain only the terms w_1, w_3 and B. Assuming for the moment that p_1 and q_1 are known, we then obtain the two equations,

$$\frac{\partial L}{\partial w_1} = \frac{a_1 \bar{w} p_1 - n_1 p_1^2 (w_1 p_1 + 2q_1)}{(w_1 p_1 + 2q_1)\bar{w}}$$

and

$$\frac{\partial L}{\partial w_3} = \frac{b_1 \bar{w} - n_1 w_3 q_1^2}{w_3 \bar{w}}$$

in which $\bar{w} = w_1 p_1^2 + 2p_1 q_1 + w_3 q_1^2$. If dominance of fitness as well as appearance is assumed, so that $w_1 = 1$, the second equation provides the estimate $w_3 = b(1 - q^2)/aq^2$, which is equivalent to the estimation of the fitness of mimics releative to non-mimics in the example on p. 187. To obtain a fitness value for each homozygote when p_1 and q_1 are unknown it is necessary to substitute the expressions in terms of B, w_1 and w_3 for p_1 and q_1. The latter frequencies are now functions of w_1 and w_3 and have to be differentiated in the expressions $\partial L/\partial w_1$, $\partial L/\partial w_3$, etc. The solution may then be found by the method of successive approximation. The result is an unwieldy set of equations, which may, however, be solved on the computer with comparatively little difficulty.

BIBLIOGRAPHY AND REFERENCES

BOOKS DEALING WITH THE MATHEMATICS OF POPULATION GENETICS

Haldane, J. B. S. 1932. *The causes of evolution*, appendix. Longmans, London; reprinted, Cornell, 1966.

Hogben, L. 1946. *An introduction to mathematical genetics*. Norton, New York.

Malécot, G. 1948. *Les mathématiques de l'hérédité*. Masson, Paris.

Li, C. C. 1955. *Population genetics*. University of Chicago Press.

Kempthorne, O. 1957. *An introduction to genetic statistics*. Wiley, New York.

Falconer, D. S. 1960. *Introduction to quantitative genetics*. Oliver and Boyd, Edinburgh.

Moran, P. A. P. 1962. *The statistical processes of evolutionary theory*. Clarendon, Oxford.

Mather, W. B. 1964. *Principles of quantitative genetics*. Burgess, Minneapolis, Minn.

Smith, J. M. 1968. *Mathematical ideas in biology*. Cambridge University Press.

Wright, S. 1968. *Evolution and genetics of population*, vol. I 'Genetic and biometric foundations'. University of Chicago Press.

Ewens, W. J. 1969. *Population genetics*. Methuen, London.

REFERENCES

Allen, J. A., and Clarke, B. C. 1968. Evidence for apostatic selection in wild passerines. *Nature, Lond.* **220**, 501–2.

Allison, A. C. 1955. Aspects of polymorphism in Man. *Cold Spring Harbor Symp. quant. Biol.* **20**, 239–51.

Allison, A. C. 1961. Abnormal haemoglobins and erythrocyte enzyme-deficiency traits. In Harrison, G. A. (Ed.), *Genetical Variation in Human populations*. Pergamon, London.

Anderson, P. K., Dunn, L. C., and Beasley, A. B. 1964. Introduction of a lethal allele into a feral house mouse population. *Amer. Nat.* **98**, 57–64.

Andrewartha, H. G., and Birch, L. C. 1954. *The distribution and abundance of animals.* University of Chicago Press.

Andrewartha, H. G. 1957. The use of conceptual models in population ecology. *Cold Spring Harbor Symp. quant. Biol.* **22**, 219–32.

Atwood, K. C., Schneider, L. K., and Ryan, F. J. 1951. Selective mechanisms in bacteria. *Cold Spring Harbor Symp. quant. Biol.* **16**, 345–55.

Bailey, N. T. J. 1951. On estimating the size of mobile populations from recapture data. *Biometrika*, **38**, 293–306.

Bailey, N. T. J. 1961. *Introduction to the mathematical theory of genetic linkage.* Clarendon, Oxford.

Band, H. T., and Ives, P. T. 1963. Genetic structure of populations I. On the nature of the genetic load in the south Amherst population of *Drosophila melanogaster. Evolution* **17**, 198–215.

Bartlett, M. S. 1960. *Stochastic population models in ecology and epidemiology.* Methuen, London.

Beale, G. H. 1954. *The genetics of 'Paramecium aurelia'.* Cambridge University Press.

Bennett, J. H. 1958. The existence and stability of selectively balanced polymorphism at a sex-linked locus. *Aust. J. Biol. Sci.* **11**, 598–602.

Beverton, J. H., and Holt, S. J. 1958. On the Dynamics of Exploited Fish Populations. *Fishery Investigations*, ser. II, vol. 19. H.M.S.O., London.

Birch, L. C. 1960. The genetic factor in population ecology. *Amer. Nat.* **94**, 5–24.

Blaylock, B. G. 1965. Chromosomal aberrations in a natural population of *Chironomus tentans* exposed to chronic low level radiation. *Evolution* **19**, 421–9.

Bodmer, W. F. 1960. The genetics of homostyly in populations of *Primula vulgaris. Phil. Trans. Roy. Soc. B.* **242**, 517–49.

Boycott, A. E., Diver, C., Garstang, S. L., and Turner, F. M. 1931. The inheritance of sinistrality in *Limnaea peregra. Phil. Trans. Roy. Soc. Lond. B.* **219**, 51–131.

Brower, L. P. 1963. The evolution of sex-limited mimicry in butterflies. *Proc. XVI Int. Congr. Zool. Washington*, vol. 4, 173–9.

Browning, T. O. 1962. The environments of animals and plants. *J. Theoret. Biol.* **2**, 63–8.

Bruck, D. 1956. Male segregation ratio advantage as a factor in maintaining lethal alleles in wild populations of house mice. *Proc. Nat. Acad. Sci. U.S.* **43**, 152–8.

Cain, A. J., and Currey, J. D. 1963a. Area effects in *Cepaea. Phil. Trans. Roy. Soc. Lond. B.* **246**, 1–81.

Cain, A. J., and Currey, J. D. 1963b. The causes of area effects. *Heredity* **18**, 467–71.

Cain, A. J., and Sheppard, P. M. 1956. Adaptive and selective value. *Amer. Nat.* **90**, 202–3.

Cannings, C., and Edwards, A. W. F. 1968. Natural selection and the de Finetti diagram. *Ann. Hum. Genet. Lond.* **31**, 421–8.

G

Carson, H. L. 1958. Response to selection under different conditions of recombination in *Drosophila. Cold Spring Harbor Symp. quant. Biol.* **23**, 291–305.

Cavalli, L. L. 1950. The analysis of selection curves. *Biometrics*, **6**, 208–20.

Ceppellini, R. 1955. Discussion of paper by A. C. Allison. *Cold Spring Harbor Symp. quant. Biol.* **20**, 252–5.

Ceppellini, R., Siniscalco, M., and Smith, C. A. B. 1955. The estimation of gene frequencies in a random-mating population. *Ann. Hum. Genet.* **20**, 97–115.

Chapman, R. N. 1931. *Animal Ecology*. McGraw-Hill, New York.

Chitty, D. 1960. Population processes in the vole and their relevance to general theory. *Can. J. Zool.* **38**, 99–113.

Chitty, D. 1967. The natural selection of self-regulating behaviour in animal populations. *Proc. Ecol. Soc. Australia* **2**, 51–78.

Clarke, B. C. 1962. Balanced polymorphism and the diversity of sympatric species. *Syst. Ass. Publ.* **4**, 47–70.

Clarke, B. 1964. Frequency-dependent selection for the dominance of rare polymorphic genes. *Evolution* **18**, 364–9.

Clarke, B. 1966. The evolution of morph-ratio clines. *Amer. Nat.* **100**, 389–402.

Clarke, B. C., and Murray, J. J. 1962. Changes in gene-frequency in *Cepaea nemoralis* (L.); the estimation of selective values. *Heredity* **17**, 467–76.

Clarke, B., and O'Donald, P. 1964. Frequency-dependent selection. *Heredity* **19**, 201–6.

Clarke, C. A., and Sheppard, P. M. 1960. The evolution of dominance under disruptive selection. *Heredity* **14**, 73–87.

Clarke, C. A., and Sheppard, P. M. 1962. Disruptive selection and its effect on a metrical character in the butterfly *Papilio dardanus. Evolution* **16**, 214–26.

Clarke, C. A., and Sheppard, P. M. 1964. Genetic control of the melanic form *insularia* of the moth *Biston betularia* (L.). *Nature. Lond.* **202**, 215–16.

Clarke, C. A., and Sheppard, P. M. 1966. A local survey of the distribution of industrial melanic forms in the moth *Biston betularia* and estimates of the selective values of these in an industrial environment. *Proc. Roy. Soc. Lond. B.* **165**, 424–39.

Cole, L. C. 1954. The population consequences of life history phenomena. *Q. Rev. Biol.* **29**, 103–37.

Cole, L. C. 1957. Sketches of general and comparative demography. *Cold Spring Harbor Symp. quant. Biol.* **22**, 1–15.

Cook, L. M., Brower, L. P., and Alcock, J. 1969. An attempt to verify mimetic advantage in a neotropical environment. *Evolution* **23**, 339–45.

Cotterman, C. W. 1953. Regular two-allele and three-allele phenotype systems. *Am. J. Human Genet.* **5**, 193–235.

Cotterman, C. W. 1954. Estimation of gene frequencies in non-experimental populations. In Kempthorne, O., Bancroft, T. A., Gowen, J. W., and Lush, J. L. (Eds.), *Statistics and Mathematics in Biology*, Iowa State College Press.

Creed, E. R. 1963. D.Phil. Thesis. Oxford University.
Creed, E. R. 1966 Geographic variation in the two-spot ladybird in England and Wales. *Heredity* **21**, 57–72.
Crisp, D. J. 1958. The spread of *Elminius modestus* Darwin in north-west Europe. *J. mar. biol. Ass. U.K.* **37**, 483–520
Crosby, J. L. 1949. Selection of an unfavourable gene complex. *Evolution* **3**, 212–30.
Crow, J., and Kimura, M. 1965. Evolution in sexual and asexual populations. *Amer. Nat.* **99**, 439–50.
Crumpacker, D. W. 1967. Genetic loads in Maize (*Zea mays* L.) and other cross fertilized plants and animals. In Dobzhansky, Th., Hecht, M. K., and Steers, W. C. (Eds.), *Evolutionary Biology*, vol. 1. North-Holland Publ. Co., Amsterdam.
Deakin, M. A. B. 1966. Sufficient conditions for genetic polymorphism. *Amer. Nat.* **100**, 690–2.
Deevey, E. S. 1947. Life tables for natural populations of animals. *Q. Rev. Biol.* **22**, 283–314.
Duncan, C. J., and Sheppard, P. M. 1965. Sensory discrimination and its role in the evolution of Batesian mimicry. *Behaviour* **24**, 269–82.
Dunn, L. C. 1957. Studies of the genetic variability in populations of wild house mice II. Analysis of eight additional alleles at locus T. *Genetics* **42**, 299–311.
Edwards, A. W. F. 1962. Genetics and the human sex ratio. *Adv. Genetics* **11**, 239–65.
Edwards, J. H. 1965. The meaning of the association between blood groups and disease. *Ann. Hum. Genet.* **29**, 77–83.
Eeckels, R., Gatti, F., and Renoirte, A. M. 1967. Abnormal distribution of haemoglobin genotypes in negro children with severe bacterial infections. *Nature, Lond.* **216**, 382.
Falconer, D. S. 1960. *Introduction to quantitative genetics*. Oliver and Boyd, Edinburgh.
Fenner, F. 1965. Myxoma virus and *Oryctolagus cuniculus:* two colonizing species. In Baker, H. G., and Stebbins, G. L. (Eds.), *The genetics of colonizing species*. Academic Press, New York.
Fisher, R. A. 1930. The genetical theory of natural selection. Clarendon, Oxford; reprinted Dover, New York, 1958.
Fisher, R. A. 1950. Gene frequencies in a cline determined by selection and diffusion. *Biometrics* **6**, 353–61.
Fisher, R. A., and Yates, F. 1963. *Statistical tables for biological, agricultural and medical research*. Oliver and Boyd, Edinburgh.
Ford, E. B. 1940. Polymorphism and taxonomy. In Huxley, Julian (Ed.), *The New Systematics*. Clarendon, Oxford.
Ford, E. B. 1964. *Ecological Genetics*. Methuen, London.
Fretter, V., and Graham, A. 1964. Reproduction. In Wilbur, K. M., and Yonge, C. M. (Eds.), *Physiology of Mollusca*, vol. I. Academic Press, New York.
Gause, G. F. 1934. *The struggle for existence*. Williams and Wilkins, Baltimore; reprinted Hafner, New York. 1964.

Gibson, J. B., and Thoday, J. M. 1964. Effects of disruptive selection. IX Low selection intensity. *Heredity* **19**, 125–30.

Goodhart, C. B. 1963. 'Area effect' and nonadaptive variation between populations of *Cepaea* (Mollusca). *Heredity* **18**, 459–66.

Greenwood, J. J. D. 1969. Apostatic selection and population density. *Heredity* **24**, 157–61.

Haldane, J. B. S. 1930. A mathematical theory of natural and artificial selection. VII Selection intensity as a function of mortality rate. *Proc. Camb. Phil. Soc.* **27**, 131–6.

Haldane, J. B. S. 1932. *The causes of evolution.* Longmans, London; reprinted, Cornell, 1966.

Haldane, J. B. S. 1948. The theory of a cline. *J. Genet.* **48**, 277–84.

Haldane, J. B. S. 1949. Disease and evolution. *La Ricerca Scientifica Suppl.* **19**, 68–76.

Haldane, J. B. S. 1953. Animal populations and their regulation. *New Biology* **15**, 9–24.

Haldane, J. B. S. 1954. The measurement of natural selection. *Proc. 9th Int. Congr. Genet.* **1**, 480–87; reprinted, in Spiess, 1962.

Haldane, J. B. S. 1956. The relation between density regulation and natural selection. *Proc. Roy. Soc. Lond. B.* **145**, 306–8.

Haldane, J. B. S. 1957a. The cost of natural selection. *J. Genet.* **55**, 511–24.

Haldane, J. B. S. 1957b. The conditions for co-adaptation in polymorphism for inversions. *J. Genet.* **55**, 218–25.

Haldane, J. B. S. and Jayakar, S. D. 1963. Polymorphism due to selection of varying direction. *J. Genet.* **58**, 237–42.

Haldane, J. B. S., and Jayakar, S. D. 1964. Equilibria under natural selection at a sex-linked locus. *J. Genet.* **59**, 29–36.

Hamilton, W. D. 1967. Extraordinary sex ratios. *Science* **156**, 477–88.

Hickey, W. A., and Craig, G. B. 1966. Genetic distortion of sex ratio in a mosquito, *Aedes aegypti. Genetics* **53**, 1177–96.

Hiraizumi, Y., Sandler, L., and Crow, J. 1960. Meiotic drive in natural populations of *D. melanogaster* III. Populational implications of the segregation-distorter locus. *Evolution* **14**, 433–44.

Holling, C. S. 1965. The functional response of predators to prey density, and its role in mimicry and population regulation. *Mem. ent. Soc. Canada* **45**, 1–60.

Hull, P. 1964. Equilibrium of gene frequency produced by partial incompatibility of offspring with dam. *Proc. Nat. Acad. Sci. U.S.* **51**, 461–4.

Hutchinson, G. E. 1948. Circular causal systems in ecology. *Ann. N.Y. Acad. Sci.* **50**, 221–46.

Inger, R. F. 1943. Further notes on differential selection of variant juvenile snakes. *Amer. Nat.* **77**, 87–90.

Ives, P. T. 1950. Mutator genes as a major cause of gene and chromosome changes in natural populations (abstract of paper). *Genetics* **35**, 672.

Kennedy, J. S. (Ed.). 1961. *Insect Polymorphism.* Royal Entomological Society, London.

Kettlewell, H. B. D., and Berry, R. J. 1961. The study of a cline. *Amathes glareosa* Esp. and its melanic *F. edda* Staud (Lep.) in Shetland. *Heredity* **16**, 403–14.

Kimura, M. 1956. A model of a genetic system which leads to closer linkage by natural selection. *Evolution* **10**, 278–87.

Kimura, M. 1958. On the change of population fitness by natural selection. *Heredity* **12**, 145–67.

Kimura, M. 1960. Optimum mutation rate and degree of dominance as determined by the principle of minimum genetic load. *J. Genet.* **21**–34.

King, J. C. 1955. Evidence for the integration of the gene pool from studies of DDT resistance in *Drosophila*. *Cold Spring Harbor Symp. quant. Biol.* **20**, 311–17.

King, J. L. 1967. Continuously distributed factors affecting fitness. *Genetics* **55**, 483–92.

Klomp, H. 1962. The influence of climate and weather on the mean density level, the fluctuations and the regulation of animal populations. *Arch. néerl Zool.* **15**, 68–109.

Kojima, K., and Yarborough, K. M. 1967. Frequency dependent selection at the Esterase 6 locus in *Drosophila melanogaster*. *Proc. Natl. Acad. Sci. U.S.* **57**, 645–9.

Lack, D. 1954. *The natural regulation of animal numbers.* Oxford University Press.

Lack, D. 1966. *Population studies of birds.* Clarendon, Oxford.

Lamotte, M. 1951. Recherches sur la structure génétique des populations naturelles de *Cepaea nemoralis* (L.). *Bull. Biol. Suppl.* **35**, 1–238.

Lees, D. R. 1968. Genetic control of the melanic form *insularia* of the Peppered Moth *Biston betularia* (L.). *Nature, Lond.* **220**, 1249–50.

Lerner, I. M. 1958. *The genetic basis of selection.* Wiley, New York.

Leslie, P. H. 1948. Some further remarks on the use of matrices in population mathematics. *Biometrika* **35**, 213–45.

Leslie, P. H., and Ranson, R. M. 1940. The mortality, fertility and rate of natural increase of the vole (*Microtus agrestis*) as observed in the laboratory. *J. Anim. Ecol.* **9**, 27–52.

Levene, H. 1953. Genetic equilibrium when more than one ecological niche is available. *Amer. Nat.* **87**, 331–3.

Levine, L., and Lascher, B. 1965. Studies on sexual selection in mice II. Reproductive competition between black and brown males. *Amer. Nat.* **99**, 67–72.

Lewontin, R. C. 1958. A general method for investigating the equilibrium of gene frequency in a population. *Genetics* **43**, 419–34.

Lewontin, R. C., and Dunn, L. C. 1960. The evolutionary dynamics of a polymorphism in the house mouse. *Genetics* **45**, 705–22.

Lewontin, R. C., and Hubby, J. L. 1966. A molecular approach to the study of genic heterozygosity in natural populations II. Amount of variation and degree of heterozygosity in natural populations of *Drosophila pseudoobscura*. *Genetics*, **54**, 595–609.

Li, C. C. 1955a. *Population Genetics.* University of Chicago Press.

Li, C. C. 1955b. The stability of an equilibrium and the average fitness of a population. *Amer. Nat.* **89**, 281–95; reprinted, in Spiess, 1962.

Li, C. C. 1963. Equilibrium under differential selection in the sexes. *Evolution* 17, 493–6.

Li, C. C. 1967. Genetic equilibrium under selection. *Biometrics* 23, 397–484.

Livingstone, F. B. 1967. *Abnormal hemoglobins in human populations.* Aldine Publishing Co., Chicago.

MacArthur, R. H., and Connell, J. H. 1966. *The biology of populations.* Wiley, New York.

Magnus, D. B. E. 1963. Sex limited mimicry II. Visual selection in the mate choice of butterflies. *Proc. XVI Int. Congr. Zool. Washington*, vol. 4, 179–83.

Malagolowkin, C., and Carvalho, G. 1961. Direct and indirect transfer of the 'sex ratio' condition in different species of *Drosophila. Genetics* 46, 1009–13.

Mandel, S. P. H. 1959. Stable equilibrium at a sex-linked locus. *Nature, Lond.* 183, 1347–8.

Mather, K. 1953. The genetical structure of populations. *Symp. Soc. exp. Biol.* 7, 66–95.

Mather, K. 1955. Polymorphism as an outcome of disruptive selection. *Evolution* 9, 52–61.

Mather, K. 1965. *Statistical analysis in biology.* Methuen, London.

Mayr, E. 1954. Change of genetic environment and evolution. In Huxley, J. S., Hardy, A. C., and Ford, E. B. (Eds.), *Evolution as a process.* Allen and Unwin, London.

Mayr, E. 1963. *Animal species and Evolution.* Harvard University Press.

Milkman, R. D. 1967. Heterosis as a major cause of heterozygosity in nature. *Genetics* 55, 493–5.

Mode, C. J. 1958. A mathmatical model for the coevolution of obligate parasites and their hosts. *Evolution* 12, 158–65.

Moment, G. B. 1962. Reflexive selection: a possible answer to an old puzzle. *Science* 136, 262–3.

Mook, J. H., Mook, L. J., and Heikens, H. S. 1960. Further evidence for the role of 'searching images' in the hunting behaviour of tit mice. *Arch. néerl. Zool.* 13, 448–65.

Moran, P. A. P. 1962. *The statistical process of evolutionary theory.* Clarendon, Oxford.

Moran, P. A. P. 1964. On the non-existence of adaptive topographies. *Ann. Hum. Genet.* 27, 283–93.

Moree, R., and King, J. R. 1961. Experimental studies on relative viability in *Drosophila melanogaster. Genetics* 46, 1735–52.

Mosimann, J. E. 1958. The evolutionary significance of rare matings in animal populations. *Evolution* 12, 246–61.

Motulsky, A. 1963. Genetic systems involved in disease susceptibility in mammals. In Schull, W. J. (Ed.), *Genetic selection in Man.* University of Michigan Press.

Mueller, H. C. 1968. Prey selection: oddity or conspicuousness? *Nature, Lond.* 217, 92.

Murray, J. 1964. Multiple mating and effective population size in *Cepaea nemoralis. Evolution* 18, 283–91.

Murray, J. J. 1966. *Cepaea nemoralis* in the Isles of Scilly. *Proc. malac. Soc. Lond.* **37**, 167–81.

Nicholson, A. J. 1933. The balance of animal populations. *J. Anim. Ecol.* **2**, 132–78.

Nicholson, A. J. 1954. The self-adjustment of populations to change. *Aust. J. Zool.* **2**, 9–65.

O'Donald, P. 1967. A general model of sexual and natural selection. *Heredity* **22**, 449–518.

O'Donald, P. 1968. Measuring the intensity of natural selection. *Nature, Lond.* **220**, 197–8.

Owen, D. F. 1963. Polymorphism and population density in the African land snail *Limicolaria martensiana*. *Science* **140**, 666–7.

Parr, M. J., Gaskell, T. J., and George, B. J. 1968. Capture-recapture methods of estimating animal numbers. *J. Biol. Educ.* **2**, 95–117.

Person, C. 1966. Genetic polymorphism in parasitic systems. *Nature, Lond.* **212**, 266–7.

Person, C., Samborski, D. T., and Rohringer, R. 1962. The gene-for-gene concept. *Nature, Lond.* **194**, 561–2.

Peters, J. A. 1959. *Classic papers in genetics*. Prentice-Hall, Engelwood Cliffs, N.J.

Pimentel, D. 1961. Animal population regulation by the genetic feed-back mechanism. *Amer. Nat.* **95**, 65–79.

Popham, E. J. 1941. The variation in the colour of certain species of *Arctocorisa* (Hemiptera, Corixidae) and its significance. *Proc. zool. Soc. Lond. A.* **111**, 135–72.

Shaw, R. F. 1959. Equilibrium for the sex ratio factor in *Drosophila pseudoobscura*. *Amer. Nat.* **93**, 385–6.

Shaw, R. F., and Mohler, J. D. 1953. The selective significance of the sex ratio. *Amer. Nat.* **87**, 337–42.

Sheppard, P. M. 1958. *Natural Selection and Heredity*. Hutchinson, London. 3rd, revised, ed. 1967.

Sheppard, P. M. 1959. The evolution of mimicry; a problem in ecology and genetics. *Cold Spring Harbor Symp. quant. Biol.* **24**, 131–40.

Sheppard, P. M., and Cook, L. M. 1962. The manifold effects of the *medionigra* gene of the moth *Panaxia dominula* and the maintenance of a polymorphism. *Heredity* **17**, 415–26.

Silliman, R. P., and Gutsell, J. S. 1958. Experimental exploitation of fish populations. *Fish. Bull. US.* **58**, 214–252.

Simpson, G. G. 1953. *The major features of evolution*. Columbia University Press, New York.

Skellam, J. G. 1955. The mathematical approach to population dynamics. In Cragg, J. B., and Pirie, N. W. (Eds.), *The Numbers of Man and Animals*. Oliver and Boyd, Edinburgh.

Slobodkin, L. B. 1962. *Growth and regulation of animal populations*. Holt, Rinehart and Winston, New York.

Smith, C. A. B. 1966. *Biomathematics* vol. 1. Griffin, London.

Smith, C. A. B. 1969. *Biomathematics* vol. 2. Griffin, London.

Smith, F. E. 1963. Population dynamics in *Daphnia magna* and a new model for population growth. *Ecology* **44**, 651–63.

Smith, J. Maynard. 1964. Group selection and kin selection. *Nature, Lond.* **201**, 1145–6.

Smith, J. Maynard. 1965. The evolution of alarm calls. *Amer. Nat.* **99**, 59–63.

Smith, J. Maynard. 1966. Sympatric speciation. *Amer. Nat.* **100**, 637–50.

Smith, J. Maynard. 1968a. *Mathematical ideas in biology.* Cambridge University Press.

Smith, J. Maynard. 1968b. Evolution in sexual and asexual populations. *Amer. Nat.* **102**, 469–73.

Smith, J. Maynard. 1968c. 'Haldane's dilemma' and the rate of evolution. *Nature, Lond.* **219**, 1114–16.

Southwood, T. R. E. 1966. *Ecological Methods.* Methuen, London.

Spiess, E. B. 1962. *Papers on animal population genetics.* Little, Brown, Boston.

Stonehouse, B. 1968. Thermoregulatory function of growing antlers. *Nature, Lond.* **218**, 870–2.

Streams, F. A., and Pimental, D. 1961. Effects of immigration on the evolution of populations. *Amer. Nat.* **95**, 201–10.

Stride, G. O. 1956. On the courtship behaviour of *Hypolimnas misippus* L. (Lepidoptera, Nymphalidae), with notes on the mimetic association with *Danaus chrysippus* L. (Lepidoptera, Danaidae). *Brit. Jour. Anim. Behav.* **4**, 52–68.

Suley, A. C. E. 1953. Genetics of *Drosophila subobscura* VIII. Studies on the mutant *grandchildless. J. Genet.* **51**, 375–405.

Sved, J. A., Reed, T. E. and Bodmer, W. F. 1967. The number of balanced polymorphisms that can be maintained in a natural population. *Genetics* **55**, 469–81.

Thoday, J. M. 1953. Components of fitness. *Symp. Soc. exp. Biol.* **7**, 96–113.

Thoday, J. M. 1963. Correlation between gene frequency and population size. *Amer. Nat.* **97**, 409–12.

Thoday, J. M., and Boam, T. B. 1959. Effects of disruptive selection II. Polymorphism and divergence without isolation. *Heredity* **13**, 205–18.

Timofeeff-Ressovsky, N. W. 1940. Mutations and geographical speciation. In Huxley, Julian (Ed.), *The New Systematics.* Clarendon, Oxford.

Tinbergen, L. 1960. The natural control of insects in pinewoods I. Factors influencing the intensity of predation by song birds. *Arch. néerl. Zool.* **13**, 265–336.

Turner, J. R. G. 1967. Mean fitness and the equilibria in multilocus polymorphisms. *Proc. Roy. Soc. Lond. B.* **169**, 31–58.

Varley, G. C. 1958. Meaning of density dependence and related terms in population dynamics. *Nature, Lond.* **181**, 1778–81.

Van Valen, L. 1965. Selection in natural populations III. Measurement and estimation. *Evolution* **19**, 514–28.

Van Valen, L., and Weiss, R. 1966. Selection in natural populations V. Indian rats (*Rattus rattus*). *Genet. Res.* **8**, 261–7.

Wallace, B. 1948. Studies on 'sex ratio' in *Drosophila pseudoobscura* I. Selection and 'Sex Ratio'. *Evolution* **2**, 189–217.

Wallace, B. 1966. On the dispersal of *Drosophila*. *Amer. Nat.* **100**, 551–63.

Wangersky, P. J., and Cunningham, W. J. 1957. Time-lag in population models. *Cold Spring Harbor Symp. quant. Biol.*, **22**, 329–38.

Watt, K. E. F. (Ed.). 1966. *Systems Analysis in Ecology*. Academic Press, New York.

Wickler, W. 1968. *Mimicry in plants and animals*. Weidenfeld and Nicolson, London.

Williams, G. C., and Williams, D. C. 1957. Natural selection of individually harmful social adaptations amongst sibs with special reference to social insects. *Evolution* **11**, 32–9.

Williamson, M. H. 1958. Selection, controlling factors and polymorphism. *Amer. Nat.* **92**, 329–35.

Williamson, M. H. 1960. On the polymorphism of the moth *Panaxia dominula* (L.). *Heredity* **15**, 139–51.

de Wit C. T. 1960. On competition. *Versl. Landbowk. Onderz. Ned.* **66(8)**, 1–82.

Wood, R. J. 1961. Biological and genetical studies on sex ratio in DDT resistant and susceptible strains of *Aedes aegypti* Linn. *Genetica Agraria* **13**, 287–307.

Woolf, B. 1955. On estimating the relation between blood group and disease. *Ann. Hum. Genet.* **19**, 251–3.

Wright, S. 1940. Breeding structure of populations in relation to speciation. *Amer. Nat.*, **74**, 232–48.

Wright, S. 1946. Isolation by distance under diverse systems of mating. *Genetics* **31**, 39–59.

Wright, S. 1949. Adaptation and selection. In Jepson, G. L., Mayr, E., and Simpson, G. G. (Eds.), *Genetics, Paleontology and Evolution*. Princeton University Press, Princeton, N.J.

Wright, S. 1965. Factor interaction and linkage in evolution. *Proc. Roy. Soc. Lond. B.*, **162**, 80–104.

Wright, S., and Dobzhansky, Th. 1946. Genetics of natural populations XII. Experimental reproduction of some of the changes caused by natural selection in certain populations of *Drosophila pseudoobscura*. *Genetics* **31**, 125–256; reprinted, in Spiess, 1962.

Wynne-Edwards, V. C. 1962. *Animal dispersion in relation to social behaviour*. Oliver and Boyd, Edinburgh.

Yasuda, N., and Kimura, M. 1968. A gene-counting method of maximum likelihood for estimating gene frequencies in ABO and ABO-like systems. *Ann. Hum. Genet.* **31**, 409–20.

Yates, F. 1958. Comments on D. V. Lindley's review of Sir Ronald A. Fisher's *Statistical methods and scientific inference*. *Heredity* **12**, 133–5.

INDEX OF PERSONS

INDEX OF SUBJECTS